Behind the Mirror

a search for a natural history of
human knowledge

Behind
the Mirror
a search for a natural history of human knowledge

Konrad Lorenz
Translated by Ronald Taylor

HBJ

A Harvest/HBJ Book
A Helen and Kurt Wolff Book
Harcourt Brace Jovanovich
New York and London

Library of Congress Cataloging in Publication Data

Lorenz, Konrad.
    Behind the mirror.

    (A Harvest/HBJ book)
    Translation of Die Rückseite des Spiegels.
    "A Helen and Kurt Wolff book."
    Bibliography: p.
    Includes index.
    1.   Psychology, Comparative.   2.   Cognition.
I.   Title.
[BF673.L6313   1978]   121   78-6031
ISBN 0-15-611776-2

First Harvest/HBJ edition 1978
A B C D E F G H I J

Contents

Translator's note

It is my pleasure to acknowledge the help of Dr Konrad Lorenz in the preparation of this translation. I am also much indebted to Dr Malcolm Dando and his wife for giving me valuable advice on matters of scientific detail.

R.T.

Epistemological prolegomena

1 *The problem*

'The corner-stone of the scientific method is the postulate that
nature is objective'. Thus writes Jacques Monod in his well-known
work *Chance and Necessity*. And he goes on: 'To be sure, neither
reason, nor logic, nor observation ... had been lacking among
Descartes's predecessors. What was still needed was the strict
investigation and analysis imposed by the postulate of objectivity.'

It is important to realize that these statements contain two
postulates, one concerned with the object of scientific research, the
other directed at the scientist himself. In the first place one
obviously has to assume the material existence of the object of
one's investigation if the investigation is to have any meaning. At
the same time, there are certain obligations incumbent on the
scientist which are far from easy to define. If they were, there
would be no need for me to write this book.

These obligations involve an attitude towards knowledge which
seems obvious to the biologist but is far from universally accepted
by philosophers and psychologists, i.e. the assumption that all
human knowledge derives from a process of interaction between
man as a physical entity, an active, perceiving subject, and the
realities of an equally physical external world, the object of man's
perception.

The somewhat confusing origin of the two words 'subject' and
'object' is worth noting. It is a mark of their imprecision that they

have exchanged meanings since the age of scholasticism. In our present usage the word 'subject' means the experiencing, thinking, feeling agent as opposed to the objects which the agent experiences, thinks and feels. Literally *subjectum* means 'what is thrown under', in the sense of a foundation on which our whole world is based. Leibniz equated the subject with *l'âme même*, 'the soul itself'. Everything we are capable of experiencing, including all we know about objective reality outside and around us, is based on the experiences of this subject, as are all our thoughts and desires. The knowledge of one's own existence expressed in the Cartesian *cogito ergo sum* remains the most certain of all knowledge, in spite of the erroneous subjective-idealist conclusions that have been drawn from it. A considerable part of this book is devoted to refuting such conclusions.

Knowing, thinking and willing, even the observation and perception preceding them, are activities. It is strange that for the most vital, most dynamic force that there is in the world we should have found no better term than 'subject', a past participle — and thus passive! How is it that from this word 'subject', which denotes the foundation of all knowledge and experience, we derive the adjective 'subjective', defined *inter alia* as 'illusory, fanciful, arbitrary, prejudiced'? And how, in what is obviously a complementary development to this devaluation of the subjective, have we arrived at our high estimation of what is popularly called 'objective' or, which is the same thing, 'corresponding to something real'?

That such assessments should have found their way into everyday language shows that there is a general attitude, clearly definable, though not usually conscious and deliberate, towards the relationship between the perceiving subject and the object of perception. We all realize that, alongside our perception of external realities, we also undergo experiences in which our own fluctuating mental states are caught up, with subjective and objective factors superimposed on each other. We have learnt to take into account, and to compensate for, the effect which our own internal physiological states have on our perception of external realities. If I go back into my house one winter's day after having been out of doors for some while, and stroke my grandson's cheek with my hand, it will feel as though he had a fever. But it never occurs to me to suspect that he is sick, since I know that my sensation is due to the temperature of my own cold hand.

This familiar sequence of events represents a good example of a process that is of fundamental importance in our knowledge of objective reality. It brings us closer to cognition of things 'as they really are' by taking into account processes and circumstances present in the observer himself. For every time we succeed in tracing an element in our experience to 'subjective' factors, and in then excluding it from the image we form of extra-subjective reality, we come a step closer to that which exists independently of our cognition.

We construct our image of 'objective' reality by making a series of steps of this kind. The world of objects, the material world of our experience, only takes shape through our eliminating the subjective and the contingent. What causes us to believe in the reality of things is in the last analysis the constancy with which certain external impressions recur in our experience, always simultaneously and always in the same regular pattern, irrespective of variations in external conditions or in our psychological disposition. It is the insusceptibility of such groups of phenomena to contingent or subjective influences that leads us to regard these phenomena as manifestations of a reality independent of all perception, as an invariable reality which we recognize as such precisely by virtue of its in-dwelling characteristics. This is why I describe the activity of abstracting constant properties by the verb 'objectivating', and its achievements by the noun 'objectivation'.

Many philosophers who do not think in biological terms labour under the illusion that by the mere exercise of the 'will to objectivity' we are in a position to rid ourselves of all personal, subjective, one-sided attitudes, prejudices, passions and so on and elevate ourselves to a position from which to utter final verdicts and universal value judgements. In order to be able to do this we need a scientific understanding of the cognitive processes in the subjective observer. The process of understanding and the characteristics of the object to be understood can only be studied simultaneously. 'The object of knowledge and the instrument of knowledge cannot legitimately be separated but must be taken together as one whole', wrote P. W. Bridgman in an article on Niels Bohr's attitude to the theory of knowledge. Monod's forceful demand for objectivity can never be fully met, but only to the extent that we succeed, as scientists, in understanding the interaction between the perceiving subject and the perceived object, between the 'instrument of knowledge' and the 'object of knowledge'.

Bridgman's clear postulate points the way to a science that seeks to understand the nature of man's cognitive functions, and in this book I want to try and show how far in this direction our present limited knowledge will allow us to go. I shall therefore consider human understanding in the same way as any other phylogenetically evolved function which serves the purposes of survival, that is, as a function of a natural physical system interacting with a physical external world.

Implied in the basic assumption that both the cognitive subject and the object of perception share the same kind of reality is a further equally important assumption, of the truth of which we are convinced. This is that everything reflected in our subjective experience is intimately bound up with, based on, and in some mysterious way identical with, physiological processes that can be objectively analysed. This attitude towards the problem of soul and body is by no means shared by all philosophers, but it seems a natural one. When someone says that his friend has just come into the room, he certainly does not mean only his friend's subjectively experiencing soul, nor his body which is accessible to physiological investigation; what he means is exactly the union of the two. It therefore strikes me as a matter of course that we should investigate both the objective physiological processes which provide men with information about the external world, and the subjective events of our own experience and knowledge. Our conviction of the unity of man as a physical entity and an experiencing subject entitles us to draw our knowledge both from physiology and from phenomenology.

An investigation of this kind must needs pursue a double aim: on the one hand it seeks to formulate a theory of human knowledge based on biological and phylogenetic information, and on the other to produce a picture of man which matches this theory of knowledge. This means making the human mind an object of scientific investigation — a venture which many thinkers would regard, if not as an out-and-out blasphemy, at least as exceeding the competence of science and as 'biologizing'. To this I would counter that no scientific understanding of man's physiological functions can in any way detract from the value of the higher activities based on these functions. I hope also to be able to demonstrate to philosophically inclined anthropologists who are suspicious of biology and phylogeny that the specifically human functions and characteristics of man reveal themselves in their uniqueness precisely when one regards them through the eyes of a scientist as the products of a natural creative process.

This call for a scientific investigation into the nature of our cognitive processes does not spring solely from an insistence on the objectivation of knowledge; it also rests on firm practical, and above all ethical, considerations.

The essence of what we call the human mind resides in a supra-individual synthesis of knowledge, volition and skill produced by the human faculty for accumulating transmitted knowledge. The resulting entity, however, is itself a living organism, synthesized from other simpler living organisms, and however supreme it may appear it shares with them an ineluctable fate — namely, that, like all living organisms, the human mind, and thus human civilization, can suffer from malfunctions. Both can become sick. So it is not only the scientist but also the doctor who, for other and more urgent reasons, calls for a scientific picture of man.

Oswald Spengler was the first to recognize that civilizations decline and perish once they have reached the point of 'high culture' in their history. He believed that there was an inescapable 'logic of the time' and an irreversible process of ageing which was responsible for the decay of all 'high cultures', including our own. But when, as an ethologist and a doctor, I observe the signs of decay in our own civilization, I cannot fail to be aware, even with the paucity of our present-day knowledge, of a series of malfunctions of an obviously pathological nature.

It is necessary to examine the ills of our civilization, not only because we hope they can be cured, but also for methodological reasons inherent in all fundamental research. For far from hindering the investigation of the organism affected by it, a pathological disorder very often gives us the key to the understanding of how that organism works. The history of medicine affords many examples of this, and in physiology the use of experimental lesions is a familiar and rewarding technique.

In my original plan, the last chapter of this book was to have been devoted to the symptoms of decay in our civilization, the 'maladies of the human mind'. But such is the transformation that has come over mankind in modern times that in the space of a few years I found myself forced to revise my assessment of the importance of these symptoms and turn this one chapter into a whole new volume, which is to follow. Part of the reason for this was also the unexpected response to my *Civilized Man's Eight Deadly Sins*.

2 'Hypothetical realism': the natural scientist's approach to theory of knowledge

The scientist sees man as a creature who owes his qualities and functions, including his highly developed powers of cognition, to evolution, that age-long process of genesis in the course of which all organisms have come to terms with external reality and, as we say, 'adapt' to it. This process is one of knowledge, for any adaptation to a particular circumstance of external reality presupposes that a measure of information about that circumstance has already been absorbed.

Similarly, anatomical development, morphogeny, produces in the organic system actual images of the outside world. The fish's motion and the shape of its fins reflect the hydrodynamic properties of water, which possess these properties irrespective of whether there are fins moving through it or not. As Goethe perceived, the eye is an image of the sun and of the physical properties of light, which are present irrespective of whether the eye is there to see the light or not. Likewise the behaviour of men and animals, in so far as it is adapted to their environment, is an image of that environment. The sense organs and central nervous system enable living organisms to acquire relevant information about the world and to use this information for their survival. Even the slipper animalcule (*paramecium*), which, when it meets an obstacle, first recoils slightly then swims on again in a random direction, 'knows' something quite literally 'objective' about its environment. *Objicere* means 'to throw towards': an object is something 'thrown in our path', something that opposes and obstructs us. The paramecium 'knows' only that the object is blocking its progress in the original direction, but as 'knowledge' this will meet any criteria which from our own infinitely more complex, more sophisticated standpoint, we are able to apply. We might often suggest directions that the paramecium could follow to better advantage than that which it haphazardly chooses, but *what* it 'knows' is absolutely correct — namely, that it cannot go straight ahead.

Everything we know about the material world in which we live derives from our phylogenetically evolved mechanisms for acquiring information, mechanisms infinitely more complex than those which elicit the avoidance response of the paramecium but developed according to the same principles. We know of nothing that can be made the object of scientific investigation but what we learn about in this way.

It follows that the human capacity for understanding reality needs to be assessed in a different way from that adopted by most epistemologists hitherto. I have only very modest hopes of even understanding the meaning or the ultimate values of our world, but I am unshakeably convinced that all the information conveyed to us by our cognitive apparatus corresponds to actual realities.

This attitude rests on the realization that our cognitive apparatus is itself an objective reality which has acquired its present form through contact with and adaptation to equally real things in the outer world. This is the basis of our conviction that whatever our cognitive faculty communicates to us corresponds to something real. The 'spectacles' of our modes of thought and perception, such as causality, substance, quality, time and place, are functions of a neurosensory organization that has evolved in the service of survival. When we look through these 'spectacles', therefore, we do not see, as transcendental idealists assume, some unpredictable distortion of reality which does not correspond in the least with things as they really are, and therefore cannot be regarded as an image of the outer world. What we experience is indeed a real image of reality — albeit an extremely simple one, only just sufficing for our own practical purposes; we have developed 'organs' only for those aspects of reality of which, in the interest of survival, it was imperative for our species to take account, so that selection pressure produced this particular cognitive apparatus. In this respect our cognitive apparatus resembles that of a primitive hunter of whales or seals, who only knows about his quarry what it is of practical value for his purposes to know. Yet what little our sense organs and nervous system have permitted us to learn has proved its value over endless years of experience, and we may trust it — as far as it goes. For we must assume that reality also has many other aspects which are not vital for us, barbaric seal hunters that we are, to know, and for which we have no 'organ', because we have not been compelled in the course of our evolution to develop means of adapting to them. We cannot hear what is transmitted on wavelengths inaccessible to our receiving apparatus, nor can we know how many such wavelengths there are.

I am a natural scientist and a doctor. At an early age I realized that in the interests of objectivity a scientist must understand the physiological and psychological mechanisms by which experiences are conveyed to men. He must understand them for the same reasons that a biologist must know his microscope and understand its optical functions — namely, in order to avoid taking for one of the characteristics of the object he is observing something that in fact results from the limitations of his instrument, such as regarding the attractive rainbow fringes caused by a not entirely achromatic lens as a feature of the particular organism he is examining. Goethe made a similar error by regarding colour qualities not as products of our perceiving apparatus but as physical properties of light itself. I shall go into this matter more fully in the section on colour constancy.

As I have said, this view of the relationship between the observer and the thing observed was formed in my young days, during the 1930s. My teacher Karl Bühler encouraged me to regard it not merely as self-evident but as the common property of all scientific thinkers, and today I am as firmly convinced of it as ever. So are others. Thus Karl Popper writes in *The Logic of Scientific Discovery* with a marked absence of any need to labour the point: 'The thing-in-itself is unknowable: we can only know its appearances, which are to be understood (as pointed out by Kant) as resulting from the thing-in-itself and from our own perceiving apparatus. Thus appearances result from a kind of interaction between the things-in-themselves and ourselves.' In his essay 'Evolutionary Epistemology' Donald T. Campbell has convincingly demonstrated why and how it is necessary for an understanding of our cognitive apparatus to know how it has evolved phylogenetically. Campbell has coined the term 'hypothetical realism' to describe this form of epistemology. It is an approach that also received the express approval of no less a man than Max Planck, who wrote to me of his deep satisfaction that, starting from such different premises, he and I should both have arrived at such similar views on the relationship between the phenomenal world and the real world.

3 *Hypothetical realism and transcendental idealism*

Almost down to the moment when in 1940 Eduard Baumgarten and I were elected, as the last of his successors, to Immanuel Kant's chair at the University of Königsberg, I held a view which Popper's statements quoted above could also tempt one to adopt — namely, that what I had called '*Weltbildapparat*' and Popper called 'the perceiving apparatus' could be unconditionally equated with Kant's concept of the '*a priori* ' .

But it is an erroneous view. There is in Kant's transcendental idealism no relationship of correspondence between the *Ding an sich* and the form in which our *a priori* 'forms of ideation' (e.g. time and space) and our *a priori* form of thought (the Kantian 'categories') make it part of our experience. What we experience is for Kant not an image of reality, not even a crude or distorted image. He saw clearly that the forms of apprehension available to us are determined by pre-existing structures of the experiencing subject and not by those of the object apprehended, but he did not see that the structure of our perceiving apparatus had anything to do with reality. In §11 of the Prolegomena to the *Critique of Pure Reason* he wrote:

> If one were to entertain the slightest doubt that space and time did not relate to the *Ding an sich* but merely to its relationship to sensuous reality, I cannot see how one can possibly affect to know, *a priori* and in advance of any empirical knowledge of things, i.e. before they are set before us, how we shall have to visualize them as we do in the case of space and time.

Kant was obviously convinced that an answer to this question in terms of natural science was categorically impossible. In the fact that our forms of ideation and categories of thought are not, as Hume and other empiricists had believed, the products of individual experience, he found clear proof that they are logically inevitable — *a priori*, and thus not 'evolved'.

What a biologist familiar with the facts of evolution would regard as the obvious answer to Kant's question was, at that time, beyond the scope of the greatest of thinkers. The simple answer is that the system of sense organs and nerves that enables living things to survive and orientate themselves in the outer world has evolved phylogenetically through confrontation with an adaptation to that form of reality which we experience as phenomenal space. This system thus exists *a priori* to the extent that it is present before the individual experiences anything, and must be present if experience is to be possible. But its function is also historically evolved and in this respect not *a priori*. The paramecium, for example, makes do, so to speak, with a one-dimensional 'ideation of space' but we cannot know how many dimensions there are to space *an sich*.

Physiological research has discovered the analysable mechanisms that govern our ideation of three-dimensional, Euclidian space. Erich von Holst investigated with great precision the functions of the sense organs and the nervous system which, on the basis of data supplied by the retina and of messages concerning the movement and focusing of the eyes, calculate the size and distance of the objects seen, thus endowing our field of vision with depth. Similarly, the messages we receive from our touch spots and our depth sensibility concerning the position of our body and limbs also communicate to us a clear image of space. The labyrinth of the ear, with its utricle and its three membranous semicircular canals, registers movement of the body in space, responding to any change of position in space, to rotation and its acceleration and deceleration. It seems to me far-fetched to assume that all these organs and functions have nothing to do with space as an *a priori* form of ideation. Rather it appears to me self-evident that they are at the root of our phenomenal three-dimensional, Euclidian space — indeed, that in a certain sense they *are* this particular form of ideation. Mathematicians tell us that other, multidimensional kinds of space are conceivable, while physicists and theorists of relativity have demonstrated that space has at least four dimensions. But we are only able to visualize that simpler version which the organization of our sense organs and nervous system enables us to apprehend.

What I have illustrated above from the relationship between the physiological apparatus with which we apprehend space and the phenomenal space in which we live, also applies *mutatis mutandis* to the relationship between all our innate forms of potential experience and the facts of objective reality which these forms of experience make it possible for us to experience. The same is true of time: here the physiologist can point to mechanisms which function like 'internal clocks', controlling our experience of the phenomenal passage of time.

Of special interest to the scientist striving for objectivation is the study of those perceptual functions which convey to us the experience of qualities constantly inherent in certain things in our environment. If, for instance, we perceive a certain object (say a sheet of paper) as 'white', even when different coloured lights, reflecting different wavelengths, are thrown on it, this so-called constancy phenomenon is achieved by the function of a highly complex physiological apparatus which computes, from the colour of the illumination and the colour reflected, the object's constantly inherent property which we call its colour.

Other neural mechanisms enable us to see that an object which we observe from various sides retains one and the same shape even though the image on our retina assumes a great variety of forms. Other mechanisms make it possible for us to apprehend that an object we observe from various distances remains the same size, although the size of the retinal image decreases with distance. The physiological functions underlying these constancy phenomena are of the greatest interest in the context of theory of knowledge because they are exactly parallel to the process of deliberate, rational objectivation referred to above. Returning to the example I gave above: as I take the temperature of my hand into account and thus make my subjective impression that my grandson has a fever more 'objective', so our 'constancy-seeking' perception of the actual colour of the object ignores the temporary colour of the illumination in order to find out the colour-reflecting properties constantly inherent in the object. These perceptual processes, which are quite beyond the reach of our powers of self-observation, also resemble those of conscious objectivation and abstraction in that they enable us to recognize certain entities in our environment as 'things' or 'objects'. That a number of physiological mechanisms have adapted themselves to this particular function serves to strengthen our conviction of the reality of the outside world. I cannot understand how one can doubt that there are objective realities behind phenomena which are consistently recorded by so many independent mechanisms, testifying to the facts like reliable, independent witnesses. As the Freiburg philosopher Szilasi used to say, in the blunt style that his limited command of the German language forced him to adopt, '*Gibt es nicht ein Ding-an-sich, gibt es viele Ding-an-siche*' ('There be not one thing-in-itself, there be many thing-in-themselves').

As it is possible to compare the various physiological mechanisms that allow human beings to experience the world around them, so it is also possible to compare the mechanisms that enable different animal species to experience the relevant facts about their environment according to their needs. These various kinds of perceiving apparatus of the manifold species inevitably differ a great deal one from the other. Not only are they differentiated according to their specific level of development and in their individual response to various aspects of their environment, but different species have a totally different 'interest' in certain aspects of objective reality. Colour constancy is important for the honey bee, which needs to be able to recognize a particular blossom by its invariable colour. For the cat, on the other hand, which hunts at dusk, colour is entirely irrelevant; what it requires is a keen perception of movement. The owl must be able to locate acoustically where the rustling of a mouse is coming from, and so on.

In the face of the immense diversity of these perceiving apparatus one fact emerges as of paramount importance, i.e. that messages pertaining to one particular environmental aspect never contradict each other. Even the avoidance reaction of the paramecium is a response to the identical reality which also appears in the incomparably more sophisticated world of human beings.

A similar example which shows that animals, with their more primitive reaction norm, evidently deal with the same objective reality as the much more sophisticated perceiving apparatus of men, is to be found in the ability to develop conditioned responses, an ability which obviously showed itself fairly early in the phylogeny of animals, and has its parallel in the specifically human concept of causality. Both can be taken as examples of how creatures have adapted themselves to the fact that all trans-formations of energy involve a specific chronological sequence of events that can be expected to follow. I shall have cause to return to the analogies between conditioned responses and causality in a later connection (see p.96 ff).

The complete agreement between the representations of the outside world provided by such an immense variety of world-depicting apparatus evolved independently by such a vast number of different forms of life certainly needs an explanation. I think it absurd to seek for any other than that all these manifold forms of possible experience refer to *one and the same real universe*. As I once said, late at night after a discussion at the Kant Society in Königsberg: 'If all of us agree that at the moment there are five wine glasses standing on the table, I fail to see how anyone in his right mind can explain this fact otherwise than by asserting that whatever "thing-in-itself" may be lurking behind the phenomenon "wine glass" is actually present five times.'

The consistent Neo-Kantian would reply that knowledge of

physical facts and acceptance of their reality are pre-conditions for our belief that we can form definite ideas about the perceiving apparatus whose physical reality we also presuppose. But, he would continue, both of these are part of our 'physical world picture', which, to the transcendental idealist, is in no way a true image of the world; to attempt to prove the one from the other would be like Baron Münchhausen pulling himself out of the mire by his own hair.

This is not a valid argument. An oversimplified presentation may give the impression that the first step in scientific investigation is to assume the existence of physical reality. If, for example, one sets out to explain the functioning of colour vision, one usually begins with the physical nature of light and the continuum of wavelengths, only then passing on to the physiological processes which turn this continuum into a discontinuum of different qualities. But this train of thought, along which the teacher is wont to lead, does not in any way correspond to the path taken by scientific discovery. This path invariably starts with subjective experience, the simple perception of colours, and then proceeds to the discovery that all the colours of the rainbow are contained in sunlight dispersed by a prism. Without the physiological mechanism which divides the continuum of different wavelengths into bands which we experience as a number of different colours, physicists would never have noticed the connection between wavelength and the angle at which the light is refracted by the prism. Once the student of objective phenomena, the physicist, had grasped this, it again became the turn of the psychologist, when Wilhelm Ostwald discovered the computing mechanism of colour constancy and its importance for the preservation of life. This showed that the alleged incompatibility of Newton's theory of colour and Goethe's is not a problem at all.

The way in which our knowledge about light and about our perception of light has developed, the one approach supporting the other, is a good example of what Bridgman meant (see p.4 f above). This is not to behave like Münchhausen, whose story of pulling himself up by his own hair is the classic allegory of the circular argument, but like a respectable pedestrian who carefully puts one foot in front of the other. The relationship between the movement first of one foot, then of the other, is what is known as the principle of mutual elucidation. If one looks first at our cognitive apparatus, then at the things which it reflects in one way or another and if on both occasions one obtains results which throw light on each other, then one can only explain this on the basis of hypothetical realism, that is, the assumption that all knowledge derives from an interaction between the perceiving subject and the object of perception, both of which are equally real. Indeed this fact justifies us in calling our epistemological approach hypothetical. This, as everybody knows, is legitimate only if an assumption can be falsified by further experience. And whenever even a small addition to our knowledge of our perceiving apparatus forces a slight correction of the picture of objective reality that it gives us, and a small step forwards in our knowledge of the essence of reality enables us to subject our perceiving apparatus to further critical examination, we become all the more justified in believing that this theory of knowledge, the naturalness of which must not be confused with naiveté, is correct.

4 *Idealism as an obstacle to knowledge*

Idealism as traditionally defined posits that the external world does not exist independently of consciousness but only as an object of potential experience. The transcendental idealism of Kant does not fall within this definition, since Kant assumed the existence of a reality *an sich* to be beyond all possible experience. Nothing of what I have to say here about the obstacle that idealism presents to knowledge refers to Kant, for Kant's postulate of the absolute unknowability of the *Ding an sich* has not prevented anybody from reflecting on the relationship between the phenomenal world and the real world. Indeed, I even have a heretical suspicion that Kant himself, less logical but far wiser in this respect than the Neo-Kantians, was in his heart of hearts not so completely convinced that the two worlds were unconnected. How otherwise could the heavens — which, by the terms of his postulate, only belonged to the world of appearances, which is indifferent to value — have repeatedly aroused the same sense of wonder in him as the moral law?

Any person not 'sicklied o'er with the pale cast' of philosophical thought will regard it as utterly perverse to believe that the everyday objects around us only become real through our experience of them. Any normal man believes that the furniture in his bedroom is still there after he has left the room. The scientist who knows about evolution is firmly convinced of the reality of the external world:

the sun shone for ages before there were eyes to see it. Whatever lies behind our ideational forms of space and the empirical principles of causality may have existed from the beginning of time. The notion that all this grandeur and probable infinity should only become reality when man, here today and gone tomorrow, happens to notice them, strikes a man of nature, be he farmer or biologist, as not only preposterous but completely blasphemous. In the face of such facts it is extraordinary that for centuries the wisest men, the greatest philosophers, above all Plato, should have been convinced idealists in the strict sense defined above.

From childhood a Platonic, idealistic attitude has been drummed into all of us, but particularly the Germans, by teachers and great poets to such an extent that we take it for granted. But when we really begin to wonder about it, we are forced to ask what reasons could have caused so many earnest seekers after knowledge to see the relationship between the phenomenal world and the real world in this topsy-turvy way.

I should like to venture an explanation of how this paradox arose. The discovery of the individual ego, the beginning of reflection, must have had a profound effect on the historical development of human thought. Not without reason has man been defined as 'the reflective creature', and the realization that man himself is the mirror in and by which reality is reflected has inevitably had considerable repercussions on all other human cognitive functions, raising them to a higher level of integration. This is also the prerequisite of the principle of objectivation which in turn forms the foundation of all science. Reflection, man's greatest discovery in the history of the human mind, was immediately followed by the greatest and gravest mistake — that of doubting the reality of the external world. Perhaps it was the very greatness of the discovery, the consternation that it caused, that made our ancestors doubt the most obvious thing of all. *Cogito*, *ergo sum*: 'I think, therefore I am' — this is certain.[1]* But who can know, or prove, that the great wide world around us is also real? To the dreamer dreams can be just as real, just as rich in experience. Is the world perhaps only a dream?

Thoughts such as these must have struck with overwhelming force the man who had just emerged from the twilight of an unreflective, 'animal' realism, and it is understandable that, beset

* For Notes, see p.246 *ff*.

by such doubts, he should turn his back on the external world and concentrate his whole attention on the newly discovered inner world. Most ancient Greek philosophers did this, and the natural sciences, which had just begun to burgeon, withered away. If a look inwards would reveal truths of the most striking profundity, while by looking outwards one could at best discover only the principles of dreams or fantasies, who would want to involve himself in laborious investigations of the external world, several aspects of which appeared far from attractive in any case?

Thus a branch of learning came into being which concerned itself almost exclusively with man as subject, with the laws of human perception, human thought and human feeling. The primacy accorded to these functions, including those of our perceiving apparatus, had the paradoxical consequence that image and reality became confused with each other — the images formed by our perceiving apparatus were taken for reality, and real objects to be imperfect, fleeting shadows of perfect imperishable ideas. *Idealia sunt realia ante rem*: the reality of the universal takes precedence over that of the particular. The 'idea of dog-ness', so charmingly ridiculed by the poet Christian Morgenstern, has for the idealist a higher reality than a living dog, or even than all living dogs.

I see the explanation for this paradox in an anthropomorphization of the creative process. When a joiner makes a table, the idea of the table is present before its 'realization' in the form of the finished product. It is also more perfect —there would be, for example, no knots in the wood —and more durable than an actual piece of furniture, for if a table is smashed, or eaten by wood-worm, repairing it or building a new one gives the idea a fresh physical form.

But a dog, like the great majority of things in our universe, is not the result of human planning. The idea of a dog that we carry in our minds is an abstraction, born, with the help of our senses and our nervous system, of our experience of numerous real dogs. We are scarcely any longer aware of the incredible paradox of taking a material object as an image of what is in fact the image of that object. The Germans, more than any other nation, are steeped in Platonic idealism. '*Alles Vergängliche ist nur ein Gleichnis*' ('All transitory things are but a symbol', wrote Goethe, Germany's greatest poet — and nobody contradicted him.[2]

Even today we are far from overcoming the obstacles which this faith in inner experience and distrust of everything external have put in the way of our search for knowledge. Until recent times all important philosophers were idealists. Modern science came into existence with Galileo, without any real help on the part of philosophy; it burgeoned from new seeds, not through a revival of the long defunct science of antiquity. It paid no heed to the discoveries of the philosophers, who in turn studiously ignored the new natural scientists. Thus the seal was set on the division between 'Art' (meaning the humanities) and 'Science'. 'Once a cultural divide gets established', said C. P. Snow in his book *The Two Cultures*, 'all the social forces operate to make it not less rigid, but more so.'

One of these social forces is mutual scorn. The Neo-Kantian Kurt Leider, for example, a colleague of mine at Königsberg, defined science as 'the acme of dogmatic narrow-mindedness', while my teacher Oskar Heinroth dismissed philosophy as 'the pathological inactivity of faculties given to man to enable him to understand nature'. Even philosophers and scientists who have a higher opinion of each other than this, amounting even to mutual respect, do not expect from the other sphere any increase in knowledge relevant to their own work, and feel no obligation even to keep themselves informed of what is happening on the other side of the fence.

In this way a barrier grew up which obstructed the progress of human knowledge in that very direction where it was most necessary, i.e. in the objective investigation of the interaction between the perceiving mind and the object of perception. For a long time the natural relationships between man and the world he lives in aroused no interest and were left unexplored.

It is to psychology that the credit belongs for making the first attempts to break down this barrier between the humanities and the sciences. The prime movers in this were the Gestalt psychologists at the turn of the century, though they were hampered at first by an inadequate knowledge of the theory of evolution. In the realm of science proper Max Planck was one of the first to attempt a breakthrough from physics, the most fundamental of the natural sciences, to epistemology, the most fundamental of philosophical disciplines. Thoroughly conversant with Kant's thought, he made the revolutionary step of treating the category of causality, which to the transcendental idealist is *a priori* and necessary, simply as a hypothesis, to be rejected when it no longer conformed to experimentally deduced facts, and replaced by the calculation of probabilities. Without his profound knowledge of Kant Max

Planck would hardly have succeeded in making this breakthrough, which is as revolutionary for the theory of knowledge as it is for physics.

A breakthrough in the other direction from philosophy to natural science seems to have been achieved by a few philosophers whose understanding of the concept of objectivation is similar to that of the scientists, and who study man from the same point of view. I am thinking here of Karl Popper, with his *Objective Knowledge* and *The Logic of Scientific Discovery* (1962), Donald Campbell's essay 'Evolutionary Epistemology' (1966) and Walter Robert Corti's 'Genetic Philosophy'.

I myself only came to realize late in life that the human mind and human civilization can, and must, be studied by using the methodology of natural science. The exclusion of all cultural considerations from my mental activities has to be seen in the light of the traditional division into sciences and humanities, since other aspects of intellectual life had interested me right from my early days. And as I have mentioned before, I recognized long ago the necessity for considering epistemology as 'knowledge of mechanisms', especially as it is even more necessary in ethology than in other branches of biology to know the limitations of one's own perceiving apparatus. So although I had an epistemological standpoint and although I was aware that man has innate norms of behaviour which are accessible to the methods and techniques of science, my research stopped short when confronted with the specific characteristics and functions of human culture.

In the end it was the physician in me that rebelled against this narrow-mindedness.[3] The progressive decay of our civilization is so obviously pathological in nature, and so obviously shows the symptoms of mental sickness, that it has become a categorical necessity to employ the diagnostic methods of medical science in order to study human civilization and the human mind. Any attempt to restore the working of a faulty mechanism must be based on an understanding of how the mechanism is constructed; there is little hope of getting it to work again without a causal understanding of both its normal functioning and the nature of the breakdown. At the same time it is worth remembering, as I mentioned above, that a breakdown often leads us to a better understanding of how the system works when it is running normally.

The majority of the psychological illnesses and disorders that threaten the survival of our civilization concern man's social and moral behaviour. To remedy this we need a scientific understanding of the causes of these pathological symptoms, and this requires us to break down the barrier between the humanities and the sciences at a point where it is stoutly defended on both sides. For as scientists refrain from passing value judgements, so students of the humanities are strongly influenced by the idealist view that anything capable of scientific explanation is *ipso facto* unconcerned with value. Thus the barrier is being strengthened at the very point where it most urgently needs to be broken down. The philosophical camp considers it blasphemy to utter the banal truth that man, like all living creatures, behaves in ways acquired phylogenetically and specified by heredity. On the other hand, many scientists often express incredulity and thinly disguised scorn when, as is proper to ethology, one begins one's investigations by observing and describing, instead of confining oneself to experimental methods and the definition of concepts in a manner which is fashionably called 'precise' and 'scientific'. It does not occur to any of these thinkers that Kepler and Newton discovered the laws that govern the movements of the heavenly bodies solely on the basis of observation and description, without making a single experiment. Still less does it occur to them that such methods might reveal the principle, far less accessible to experiment than the law of gravitation, which governs our own ethical and moral behaviour. So man's path to self-knowledge remains firmly blocked. A few — all too few — are trying to remove the obstacles; indeed, their number is gradually growing, as is also their devotion to the task, convinced, as they are, that the fortunes of mankind depend on the success of their efforts. There is no doubt that truth will ultimately prevail. But the worrying question remains: will the victory come in time?

The realist persists in looking outwards only, unaware that he is a mirror. The idealist persists in only looking into this mirror, averting his eyes from the external world. Thus, both are inhibited from seeing that there is an obverse to every mirror. But the obverse does not reflect, and to this extent the mirror is in the same category as the objects that it reflects. The physiological mechanism whose function it is to understand the real world is no less real than the world itself. This very obverse is the subject of my book.

Chapter one

Life as a process of learning

1.1 *Positive feedback of energy gain*

The most amazing function of the life process, and also the one that is in most need of explanation, is that, in apparent contradiction to the laws of probability, it seems to develop from the simple to the complex, from the more probable to the less probable, from systems of a lower order to systems of higher order. However, none of the laws of physics, even the second law of thermodynamics, is broken by this. All life processes are sustained by the flow of energy being 'dissipated' — as physicists say — in the universe. Or, as a Viennese friend of mine once put it, 'Life feeds on negative entropy.'

All living systems are constructed in such a way as to be able to acquire and store energy. Otto Rössler drew an attractive analogy between life sustained by the dissipation of energy in the world and a sandbank formed across a river sideways on to the current: the more sand has piled up, the more the sandbank is able to collect. The more energy living systems have absorbed, the more they will be able to absorb: when a living organism flourishes, it grows and propagates. A large number of big animals will eat more than a small number of little ones. Organisms are thus systems which derive their energy in this cycle of so-called 'positive feedback'.

One finds similar systems in the inorganic world. The term 'snowballing' denotes the same process: an avalanche gets larger and larger; the bigger a fire, the more rapidly it spreads, and so on. The flame occurs often in poetry and in proverbs as a symbol of life and growth.

1.2 *Adaptation as acquisition of knowledge*

Organic systems differ from these inorganic systems in one vital respect: they owe their ability to acquire and store energy to certain often highly complex physical structures, which, by a process which makes them capable of performing this function, have been phylogenetically developed for the purpose.

Thanks to the discoveries of Darwin and recent research in biochemistry, we are now able to form clear pictures of the processes by which the adaptiveness of organic structures comes about. The structural plan of every species of living organism is laid down in the two strands of chain-like nucleic acid molecules, the sequence of the nucleotides forming a code. This code is replicated at each mitotic division of the cell, the double chain of the molecule splitting in half and each half becoming a double chain again by 'collecting' free nucleotides and joining them to itself in the same order as that of the half that was separated off. Thus new double chains are formed, consisting of the old half and a new, complementary half. Genetic continuity therefore depends on material continuity, with the reservation, however, as Weidel put it, 'that what is passed from generation to generation is a certain *structure* of matter'. Sometimes slight errors occur in the replication of the chains, with the result that the code of the newly formed molecular strand deviates in certain details from the previous one. This is what is known as genetic mutation.

In all organisms with proper nuclei, the so-called cariotes, which include all higher animals and plants, the genes are grouped together in larger units called chromosomes, which occur in pairs in every nucleus. In each pair of chromosomes there are identical or corresponding genes, grouped in approximately the same order. Before sexual reproduction each pair of chromosomes divides, leaving the reproductive cells with only half a set of chromosomes, which is called the haploid state of the cell; on fertilization the chromosomes form new pairs of which one element derives from the paternal and one from the maternal side; in this way, as well as through mutation in the chromosomes, new combinations of genetic factors can be produced. As a result of these processes, described here in a very simplified way, the external appearance of higher organisms, the so-called phenotype, is never entirely constant.

The extent and frequency of these variations are such that they do not threaten the survival of the species by the production of unviable aberrations, but they certainly do not always work to the advantage of the individual organisms. Indeed, since all these slight, sometimes infinitesimal, variations are entirely random, their most frequent consequence is to reduce the prospects of energy gain and survival. Only in exceptional cases — but it is these that matter here — does a mutation or new combination of genetic factors enable an organism to adapt more profitably to its environment than its ancestors had done. In such cases the new organism is better fitted to one or the other element of its environment, thus gaining a better chance to acquire energy, or reducing the probability of loss of energy. The chances of survival and reproduction of the favoured organism increase at the expense of, and in competition with, its not similarly favoured fellows, which are thus doomed to extinction. This process is known as natural selection, and the modifications brought about in the organism as adaptation.

Their insight into these two processes has forced biologists to form two concepts unknown to physicists or chemists. The first is that of 'fitness for survival' or teleonomy. As natural selection 'breeds' structures which fulfil particularly well a certain survival function, they finally look as though they had been created for the very purpose by a wise and beneficent spirit. One may remark, in parenthesis, that this is not such an entirely wrong impression, for, as I shall show later in this chapter, the constructive power of the human mind derives from processes which are basically similar to those that occur in the genome.

In the last analysis all complex structures of the entire range of organisms have arisen through the selection pressure of survival functions. When a biologist comes across a structure whose function he does not know he naturally asks what purpose it serves. To ask, for example, 'Why does a cat have curved, pointed claws?' and to receive the answer, 'In order to catch mice', epitomizes the way in which a problem is posed and solved. Colin Pittendrigh coined the word 'teleonomic' for this question of survival value, seeking to draw as sharp a distinction between teleonomy and teleology as there is between astronomy and astrology.

The second concept that emerges from this process of adaptation is that of knowledge or 'information'. The word 'adapt' itself implies a process through which a correspondence is brought about between the object that is being adapted and the circumstances to which it is being adapted. What an organism learns in this way about external reality is, quite literally, 'impressed' or 'imprinted' on it.

The word information primarily means 'giving form'.

To use the word 'information' in the everyday sense could lead to misunderstanding, because information theorists give it a far wider meaning, deliberately avoiding the semantic level, e.g. any question of the content of the information, and particularly any relevance it may have for the biological survival of the organism. Thus in the terminology of information theory we cannot talk, as we do in everyday language, of 'information about something'. Common parlance, however, has two plain words for this kind of acquiring and possessing information: apprehending and knowing. However, when I speak of information as being the root of all processes of adaptation, I am using the word in its everyday sense to denote something that has a meaning and a purpose for whoever receives or possesses it.

For the benefit of the reader interested in information theory I might add that, as Bernhard Hassenstein has shown, it would be possible also to define this conception of the word 'information' in terms of information theory itself. One would then have to say that adaptation is an increment in the transinformation between the organism and its environment, the increment being caused by processes within the organism, without any perceptible change in the environment. This could be regarded as a special case of the emergence of what Meyer-Eppler called a 'correspondence'* — an asymmetrical or one-sided correspondence, however, inasmuch as it is achieved entirely by changes in only one of the two parties. As the reader can see, the terminology of information theory is hardly suited to describing life processes.

The gain in knowledge, achieved by trial and error by the genome — which retains what is fittest for it — results, as mentioned above (p.6) in the formation of an image of the material world within the living system; an image 'built up', as Donald M. Mackay puts it, from the 'descriptive information content of the situation'. The image thus formed is, in a way, a negative of reality, like a photographic negative, or the plaster cast of a coin. Uexküll described the organism as having a 'contrapuntal' relationship to its environment. As already mentioned, there are relationships of this kind between organism and reality in the field of anatomy, such as the sun-like nature of the eye, or the wave-like movements of a fish's fins. Structures of this kind owe their remarkable efficiency to their 'descriptive information content' and serve the

*Quoted from Norbert Bischof.

organism's economy to a marvellous degree, enabling it to tap the most unlikely and inaccessible sources of energy.

The method of the genome, perpetually making experiments, matching their results against reality, and retaining what is fittest, differs from that adopted by man in his scientific quest for knowledge in only one respect, and that not a vital one, namely that the genome learns only from its successes, whereas man learns also from his failures. Otherwise, like the genome, man proceeds by comparing an idea in his mind, a hypothesis he has evolved, with the outside world and 'checks whether it fits'. In his treatise *Pattern Matching as an Essential in Distal Knowing* Donald T. Campbell has convincingly shown that the greater part of all knowledge, from the simple recognition of an object to the verification of a scientific hypothesis, is arrived at in this way.

This process of 'pattern matching' can only function with groups or configurations of sense data and the relationship between them. The isolated report of the sensory cell is, in principle, always ambiguous. One cannot identify an individual star that shines through a small gap in the clouds; only when there is a larger area of clear sky visible with a number of stars in it is one in a position to compare the pattern one sees with the stellar pattern one knows, and thus identify the star one saw first, provided that it is a fixed star. If it is a planet a great deal more knowledge of the heavenly bodies and the temporal pattern of their distribution is needed to identify it.

In his 'Evolutionary Epistemology' Campbell writes: 'The natural selection paradigm of such knowledge increments can be generalized to other epistemic activities, such as learning, thought and science.' I not only agree with this statement, but I regard it as one of the main tasks of my book to undertake just such a comparison of the various mechanisms by means of which different living systems acquire and store the information relevant to their needs. Most of what science has brought to light about the external world, as Campbell rightly says, has been discovered by 'pattern matching', and since cognitive processes, some on the highest and some on the lowest level, are based on the same principle, one might conclude that there was no other way to acquire knowledge.

This erroneous view — as we shall see it to be — might appear to be strengthened by the fact that all cognitive mechanisms, the oldest and simplest as well as the most recent and most complex, have a further important characteristic in common — namely, that both the mechanism by which the genome acquires knowledge, and that by which man does the same, change with each item of knowledge acquired. Having acquired new information, neither is the same as it was before. Each new acquisition improves its chances of gaining energy and thereby the probability of acquiring further knowledge.

1.3 *The acquisition of instantaneous information not intended for storage*

But there are also cognitive functions of a quite different kind. As adaptation has created physical structures equipped to absorb and utilize energy, so it has also produced structures whose function is to acquire and exploit knowledge about circumstances which arise suddenly, without notice, in the animal's environment, and which must be instantly taken into account by its behaviour. Behaviour that depends on the operation of these mechanisms takes the form of a relevant response to a particular environmental situation, although this situation has never before arisen in this particular form either in the phylogeny of the species or in the life of the individual organism itself. Significantly, this definition is also applicable to so-called 'insight-controlled behaviour', and is as true of the simplest taxes or orientation responses as it is of the sophisticated operations of the sense organs and nervous system which lie at the root of the *a priori* forms of human intuition and thought.

The subjective phenomenon of insight, called by Karl Bühler the 'Aha experience', manifests itself in the same way whether we have just grasped the significance of a set of highly complex relationships, or whether through the function of orientation responses the state of disorientation gives way to one of blessed orientatedness, such as when the statolithic apparatus of our inner ear tells us that 'upwards' is a different direction from what it was a moment earlier. I once experienced for myself how powerful this insight can be, when, as I was lying fast asleep in a motor boat on the Danube one night, a friend tipped me overboard. So muddy was the water that even at a shallow depth there was no light to show me which way was down and which way was up. I can assure my readers that it was a traumatic experience of the meaning of 'insight' when after a few anxious moments of total disorientation, the statoliths did what was expected of them.

The processes described here are not adaptation processes like those discussed earlier (p.21), but the operations of anatomical, sensory and nervous structures whose adaptation is already completed. Those structures are as little, or even less, open to individual modification than those serving the acquisition not of information but of energy. Also, the repeated function of acquiring instantaneous information must again leave no traces in the physiological mechanism performing it. The basic function of keeping the organism informed of rapid change in its environment presupposes that the underlying apparatus remains capable of cancelling one message and substituting another, which may often be just the opposite.

There is a further and even more important consideration. The organizations which are immune to all changes, i.e. the mechanisms which, on the basis of momentary sensory messages, provide us with our immediate 'insights' into the world around us, are the foundation of all experience. They precede all experience, and must do so, if experience is to be possible at all. In this respect they correspond absolutely to Kant's definition of the *a priori*.

As we shall discuss in a variety of contexts, the reliable performance of a completely adapted structure must invariably be paid for by the loss of certain degrees of freedom. The mechanisms of these instantaneous cognitive activities are no exception. By being specifically adapted to acquire a particular kind of information, most structures are tied to a very narrow, rigid programme, their inbuilt computing mechanisms containing 'hypotheses' to which they blindly adhere. If circumstances arise that were not 'foreseen' by the adaptive process that produced them, a structure may transmit false information which it cannot be taught to correct. The various different kinds of sensory illusion provide ample proof of this.

This 'doctrinaire' element resulting from the completion of adaptive processes forces our cognitive faculties to accept various hypotheses — or rather, it foists hypotheses upon them without our noticing it. We cannot experience, perceive or think anything except on the basis of assumptions and suppositions of which such hypotheses are innate parts. They are built into our perceiving apparatus. And however hard we try, in a spirit of free inquiry, to discover new hypotheses, we cannot banish from them these ancient *a priori* hypotheses, which have evolved through mutations and new combinations of genes and been tested over countless ages of 'pattern matching', and which, though never absolutely pointless, are always rigid and never entirely relevant.

1.4 *Double feedback of energy and information*

The acquisition and storage of relevant information is as basic a function of all living organisms as is the absorption and storing of energy. Both are as old as life itself. Otto Rössler, I believe, was the first biologist to point out that not only do the energy-absorbing processes form a positive feedback cycle in themselves but they also have a positive feedback relationship to the information-acquiring processes.

When, through mutation of a new combination of genetic factors, the probability of absorbing energy increases to the point where natural selection favours the development of the new, improved organism, the number of its descendants also increases. At the same time it also becomes more probable that it will be these descendants who will win the next big prize in the mutation stakes.

This dual feedback cycle characterizes all life, including that of viruses, which, in Weidel's striking phrase, have only 'a borrowed life'. It is undeniably true, yet at the same time misleading, to say that living organisms are at the mercy of purely random changes and that evolution only takes place through the elimination of the unfit.

A closer approximation to what really happens in organic nature would be to describe it as follows: life is an eminently active enterprise aimed at acquiring both a fund of energy and a stock of knowledge, the possession of one being instrumental to the acquisition of the other. The immense effectiveness of these two feedback cycles, coupled in multiplying interaction, is the precondition, indeed the explanation, for the fact that life has the power to assert itself against the superior strength of the pitiless inorganic world, and also for the fact that it tends at times to an excessive expansion. The process whereby a large modern industrial company, such as a chemical firm, invests a considerable part of its profits in its laboratories in order to promote new discoveries and thus new sources of profit is not so much a model as a specific case of the process that is going on in all living systems.

It was an important discovery on the part of Otto Rössler that the phylogeny of the organic world does not depend on 'chance' or 'luck' but that organisms instantly seize on any favourable circumstances that come their way, thereby encouraging the incidence of further such circumstances. If we understand this, we shall come closer to a solution of two great mysteries.

The first is that of the speed of evolution. If this depended simply on the random elimination of the unfit, and if there were no feedback processes, then the period of a few thousand million years which has been calculated by physicists, on the basis of the rate of decay of radioactive substances, to be the age of our planet would hardly be long enough for man to have evolved from the most primitive organisms.

The second question is that of the *direction* of evolution. Life is both acquisition of information, i.e. a cognitive process, and economic enterprise (one is almost tempted to call it commercial). An increase in knowledge about the outside world produces economic advantages; these then exert the selection pressure which causes the mechanisms that acquire and store information to develop further.

When scientists for whom the acquisition of knowledge has become an end in itself, together with philosophers and the *homo ethicus* of civilized society in general, are confronted with the sum of potential energy on one side and the accumulated wealth of knowledge on the other, they cannot help rating the latter far higher than the former. Nor will this assessment be affected by our knowledge that the selection pressure which has created that wealth of knowledge has been economic in nature. Indeed, so universal is the pressure towards increasing the stock of information relevant to survival, that it could very well suffice in itself to explain the general direction of evolution from 'lower' to 'higher' states. I do not maintain that all other facts, some of them possibly unknown, are to be excluded, but as our present knowledge stands, there is no necessity to postulate non-natural forces like J. G. Bennett's 'Demiurgic Intelligence' in order to explain the general direction of evolution. When we use the terms 'higher' and 'lower' of living creatures and cultures alike, our evaluation refers directly to the amount of knowledge, conscious or unconscious, inherent in these living systems, irrespective of whether it has been acquired by natural selection, learning or exploratory investigation, and whether it is preserved in the genome, in the individual's memory, or in the traditions of a culture.

Chapter two

The creation of new system characteristics

2.1 *Inadequacy of vocabulary*

When one attempts to describe the great process of organic growth, one finds oneself hampered by the fact that the language of culture was born at a time when ontogeny, i.e. the evolution of the individual creature, was the only form of development known. Words like development and evolution have the etymological connotation of the unfolding of something that was already there in a compressed or confined form, like the flower in the bud, or the chicken in the egg. For ontogenic processes of this kind such words are perfectly suitable. But they are lamentably inadequate when one attempts to define the nature of an organic creative process through which something entirely new comes into existence, something that was simply not there before.

2.2 Fulguratio, *or the 'creative flash'*

Theistic philosophers and mystics of the Middle Ages coined the term *fulguratio*, 'flash of lightning', to denote the act of creation, thereby conveying the notion of a sudden intervention from above, from God. By an etymological accident or perhaps through deeper, unsuspected associations, this term is far more appropriate than those mentioned above for designating the coming into existence of something previously not there. A thunderbolt from Zeus is for the scientist an electric spark like any other, and if we see a spark at an unexpected point in a system, the first thing we think of is a short circuit, a new connection.

If, for example, two independent systems are coupled together (see Fig. 1, taken from Bernhard Hassenstein), entirely new, unexpected system characteristics will emerge, of whose appearance there was previously not the slightest suggestion. This is the profound truth behind the Gestalt psychologists' principle, mystical in tone but absolutely correct, that the whole is more than the sum of its parts.

There are many ways, as we shall see, in which such new characteristics can arise. One particular way is the following. In a series of subsystems causally linked in a linear chain, with the first functioning only as a cause and the last only as an effect, this last can, by the emergence of a new causal link, influence the first and thus turn the causal chain into a cycle, or loop. We have already met examples of such cycles with positive feedback in our discussion of how energy and information are acquired. Equally important are negative feedback loops, which I shall discuss in later chapters, where they logically belong. Suffice it for the moment to say that, if at some point in a causal cycle a 'negative sign' is built in, i.e. if the effect of one process in the chain becomes weaker as a result of the preceding process becoming stronger, the result is one of regulation. For example, the higher the water rises in a lavatory cistern, the higher it lifts the ball-cock, which then prevents more water from running in. The result is a constancy in the water level.

Cybernetics and systems theory have shown that the sudden emergence of new system characteristics has nothing to do with miracles, thereby absolving phylogeneticists from the reproach of vitalism. There is nothing supernatural about a linear causal chain joining up to form a cycle, thus producing a system whose functional properties differ fundamentally from those of all preceding systems. If an event of this kind is phylogenetically unique it may be epoch-making in the literal sense of the word. Some geological epochs have derived their names from such events.

2.3 *Unity out of diversity*

Many thinkers, philosophers and scientists alike, have recognized that progress in organic development is almost always achieved through the integration of a number of different and independent

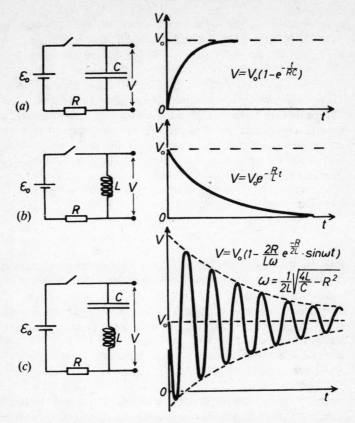

Figure 1 Three electric circuits, including an oscillating circuit (*c*). A current is passed between the poles of a battery with electromotive strength E_o or terminal voltage V_o. R = resistance. Circuit (*a*) has a condenser with capacity C; (*b*) has a coil with inductance L; circuit (*c*) has both condenser and coil. The voltage V can be measured at the two terminals.

The diagrams on the right show the changes in voltage when the switch is closed at time zero. In circuit (*a*) the condenser gradually charges through the resistance until voltage V_o is reached. In circuit (*b*) the current, initially impeded by self-induction, increases until it reaches the strength laid down by Ohm's law; the voltage V is then theoretically zero, because the total resistance is concentrated in R. In circuit (*c*) diminishing oscillations are produced. It is clear at a glance that the behaviour of (*c*) is not the result of superimposing (*a*) and (*b*) although (*c*) can be thought of as having arisen through the addition of (*a*) and (*b*). This pattern is valid, for example, for the following values: $C = 0 \cdot 7 \times 10^{-9}$ F; $L = 2 \times 10^{-3}$ Hy; $R = 10^3 \omega$; $\lambda \approx 1 \cdot 2 \times 10^{-6}$ sec. This last value also defines the time axis, which is the same for all three curves. (Calculations by E. U. von Weizsäcker.)

systems to form a unit of a higher order, the systems becoming modified in the process, the better to equip them for their roles in the new, more advanced system. Goethe defined development as differentiation and subordination of the parts. Ludwig von Bertalanffy has described this process with great precision in his *Theoretische Biologie*, while in *Science, Man and Morals* W. H. Thorpe demonstrated that the most important creative principle in evolution is the emergence of a totality from a mass of different parts which become more and more dissimilar in the process — as well as more dependent on each other. Teilhard de Chardin gave the most succinct and poetic definition: '*Créer, c'est unir*' ('To create is to unify'). This is a principle that must have been at work from the very beginning of life.

This creative union of diverse parts to form a new entity in itself implies a complication of the living system. However, the new system often becomes simplified in the course of further evolution through the 'specialization' of the individual parts — that is, through their restriction to particular functions; the other functions which the individual part had to perform while it was still independent are now assigned to other parts of the new system. Even the ganglion cells of our brain, which together perform the highest intellectual functions, are individually far inferior to an amoeba or a paramecium, both in its individual function as a cell and as concerns the relevant information that underlies this function. An amoeba or a paramecium has at its command a whole series of appropriate responses to external stimuli and 'knows' a great number of important things about its environment, whereas the ganglion cell only 'knows' when to 'fire'. Even this it cannot do with varying strength: it either does it or not, following the law of 'all or nothing'. This 'stultification' of an element that has become part of a higher entity has, of course, its good side as well, in that it is indispensable for the operation of the entity as such by guaranteeing the unequivocal transmission of messages. The strength of the message passed on by the cell must not be allowed to depend on the condition in which the cell happens to be, any more than a well-disciplined soldier can be left to decide whether to carry out an order enthusiastically or not.

This process of simplification in the course of integration into a higher entity is a phenomenon found at all stages of evolution. In the psychosocial development of man and his culture it presents special problems. The inevitable development of the practice of the division of labour leads in all areas, but most tragically in the intellectual sphere, to ever increasing specialization. The specialist comes to know more and more about less and less, until finally he knows everything about a mere nothing. There is a serious danger that the specialist, forced to compete with his colleagues in acquiring more and more pieces of more and more specialized knowledge, will become more and more ignorant about other branches of knowledge, until finally he is utterly incapable of forming any judgement on the role and importance of his own sphere within the context of human knowledge and culture as a whole.

Another form of simplification is what is called in social contexts 'improved organization'. The first, experimental forms of a new machine are always more complicated than the final version, and the same is frequently true of living systems. Interaction between the various elements, in particular the exchange of information, becomes simplified or is made more direct, and the superfluous remains of earlier stages are removed, or 'become rudimentary', as biologists say.

2.4 *The one-sided relationship between levels of integration*

The manner described above in which smaller systems become integrated to form a new and more advanced system leads to a peculiar kind of one-way relationship, both between the whole and its constituent parts within the organism, and between higher animals and their primitive ancestors. In principle the same relationship also exists between the whole of organic life and the inanimate matter from which it is derived. In ontological terms one can define the relationship by saying that the whole is its parts, and continues to be so even if, as the result of a series of *fulgurationes*, or 'creative flashes', it acquires a number of additional system characteristics in the course of its evolution. The subsidiary systems themselves do not gain any higher characteristics and may even lose some in the process of simplification. None of the laws that govern the subsidiary systems ever suffers any infraction in the new entity, least of all those of inanimate matter, out of whose elements all living systems are built.

The one-way relationship consists on the one hand in the system as a whole possessing all the characteristics, in particular all the weaknesses, of its component parts — for a chain is only as strong as its weakest link. On the other hand, none of these parts possesses the characteristics specific of the whole. Similarly, every higher organism possesses most of the characteristics of its primitive ancestors, but even the fullest knowledge of an animal's characteristics will not allow us to predict those of its more highly developed descendants.

This does not mean, of course, that highly developed systems are beyond analysis or natural explanation. But the scientist must never forget that the laws and characteristics of any system, like those of the individual subsidiary systems within it, have to be explained on the basis of the laws and characteristics of the systems on the next lower level of integration. And this, in its turn, is only possible when one is familiar with the structure of the higher entity formed by the coalescence of the systems on this level. If one assumes complete knowledge of this structure, it is theoretically possible to explain every living system and all its functions, even the most advanced, in natural terms, i.e. without adducing supernatural factors.

2.5 *The irrational residue*

However, to claim that a creature is in principle capable of being explained is a valid assertion only if we accept the present structures in its body as our data — in other words, if we maintain that we are not interested in their historical evolution. For, if we ask why a particular organism is structured in one way rather than another, the most important answers will be found in the history of the species concerned. To the question why our ears are situated on each side of our heads, the causal answer is that we are descended from water-breathing ancestors who had at these points a gill slit, the spiraculum, which was retained as an air-conducting passage during the transition to terrestrial life and then, through functional change, became part of the auditory system.

The number of purely historical causes one would need to know in order to fully explain why an organism is as it is may not be infinite but it is sufficiently great to make it impossible ever to trace all causal chains to their end, even supposing they had one. Thus we are always left with what Max Hartmann called an irrational or non-rationalizable residue. That evolution produced oak trees and human beings in the Old World but eucalyptus trees and kangaroos in Australia is simply a result of such undetectable causal sequences, which, in a mood of resignation, we have come to describe as 'chance'.

Although, as we must always emphasize, we scientists cannot believe in miracles — that is, in violations of the universal laws of nature — we are at the same time aware that we can never succeed in giving a complete explanation of how a creature has evolved from its lower ancestors. A higher animal, as Michael Polanyi in particular has pointed out, cannot be 'reduced' to its simpler ancestors; still less can a living system be 'reduced' to inorganic matter and the processes that take place within it.

The same is even true of man-made machines. From the point of view of their material construction they are completely analysable, so that they can be readily manufactured. If, on the other hand, one considers their historical, teleonomic evolution as organs of man, any attempt to explain why they are as they are, and not otherwise, brings one up against the very same non-rationalizable remainder as with living systems.

Polanyi will scarcely have intended to postulate the presence of vitalist factors, but in order to remove any possibility of misunderstanding, I would prefer to say that no system on a higher level of integration can be deduced from a lower system, however fully one may understand this lower system. We know with certainty that higher systems have arisen from lower ones, absorbing them and containing them like bricks in a building. We also know, with absolute certainty, the earlier stages in development from which higher living beings emerged. But each step forward has consisted of a *fulguratio*, a historically unique event in phylogeny which has always had a chance quality about it — the quality, one might say, of something invented.

Chapter three

Strata of existence

3.1 *Nicolai Hartmann's 'categories of existence'*

Is it really true that Kant's philosophy, as has been said, completely ignores the question of the foundation of Existence? Is it not rather the case that the problem of the distinction Kant draws, like the problem of objective validity in general, actually *raises* the question of the foundation of Existence? A rational concept can only hold true if what it says of a thing is present in the thing itself: the notion of 'objective validity', as far as it applies, must thus assume that the cognitive category also is an objective category.

As these sentences from Nicolai Hartmann show, his fundamental conviction of the existence of an objective world leads him to equate the categories of human thought with those of extra-subjective reality. 'Category' for him means 'statement', 'predicate': 'Categories are the basic predicates of Actual Things, preceding, and providing a context for, all specific predicates'. And again:

As the most general forms of statement — the guidelines, as it were, for any subsequent, more specific statements — categories predicate the basic characteristics of the objects with which they are concerned. This means that these basic characteristics belong to the objects as Actual Things, independent of whether they are predicated by them or not. All Actual Things, when expressed, take the form of predicates. But the predicates are not identical with the Things. Concepts of judgements are not there for their own sake but for the sake of the Things. It is the inner, ontological meaning of judgement that transcends its logical form and this, in spite of misunderstandings, is what has given the concept of 'category' its ontological justification.

If, as is abundantly clear from these quotations, Hartmann takes it for granted that the 'rational' category is at the same time an 'objective' category; and if, as he does, he makes this the basis of his belief in the existence and the relative knowability of the outside world, then his fundamental epistemological position is very closely related to that of hypothetical realism, in which the categories and modes of perception of man's cognitive apparatus are the natural products of phylogeny and thus adapted to the parameters of external reality in the same way, and for the same reasons, as the horse's hooves are adapted to the prairie, or the fish's fins to the water. It was hardly Hartmann's intention to look for a historico-genetic explanation of the correspondence he claims exists between the two different categories. Yet his views on the structure of the material world, especially the world of organisms, correspond so exactly to those of the phylogeneticist, that I always find it difficult to give an account of his ideas without smuggling evolutionist interpretations into his theory of 'strata' of physical life. I once asked my friend Walter Robert Corti, who knew him well, what Hartmann would have said to an evolutionary interpretation of his ideas. Corti thought he would have rejected it, then added, to console me: 'But that is what makes his ideas palatable.' So at least one real philosopher shares my opinions.

3.2 *Hartmann's theory of 'strata of existence'*

The world in which we live, says Hartmann, is built up of strata each with its own existential categories or sets of categories which distinguish it from other strata. 'There are in the hierarchy of existence certain phenomenal realities whose fundamental differences our mind fails to bridge. ... Any true and accurate theory of categories must have as much regard to these gaps as to the existential relationships that bridge them.' These relationships, however, only transcend in a unilateral manner the divisions between the four great strata of Existence — the Inorganic, the Organic, the Conscious and the Spiritual — in a unilateral manner. The principles of existence and laws of nature which govern inorganic matter apply with equal validity to the higher strata. Hartmann puts it as follows:

> Organic nature rises above inorganic nature — not freely and independently, however, but on the basis of the laws and circumstances of the world of matter, even though these laws and circumstances are far from sufficient in themselves to constitute life. Likewise spiritual life and consciousness have as their prerequisite an organism in which and with which they come into the world. In like manner the great historical moments in cultural, spiritual life are carried by the conscious life of the individuals involved. From stratum to stratum, passing across the successive divisions, we find everywhere this principle of dependence on the stratum below, yet at the same time the autonomy of the superimposed stratum with its own structures and its own laws.
>
> This relationship constitutes the real unity of the actual world. For all its variety and heterogeneity, the world is still an entity. It has the unity of a system — a system of different strata.
>
> The vital point is not that the differences between these levels are unbridgeable — indeed, it may be only to us that they appear unbridgeable — but that new laws and categories are established which, though dependent on those of the strata below, have their own character and assert their own autonomy.

These statements of Hartmann's clearly reveal the fundamental compatibility of his own views, based solely on an ontological argument, with those of the phylogeneticist who derives his knowledge from the comparative study of living creatures. Hartmann's philosophy has been rejected as being a 'pseudometaphysical construct', but this is completely wrong. It is not built on deductive speculation but on empirical evidence, and takes full account of the multifarious phenomena in the world without breaking it into a mass of heterogeneous parts.

For me the most convincing proof of the correctness of Hartmann's views is that, although they take not the slightest account of the facts of evolution, they tally absolutely with these facts, just as any sound system of comparative anatomy does, even if it belongs to pre-Darwinian days. Hartmann's sequence of strata corresponds directly to the pattern of evolution: there was inorganic matter on earth long before organic life; much later central nervous systems with a capacity for subjective experience, a 'soul', evolved; finally, only in the most recent phase of creation did the spiritual life of human culture make its appearance.

Hartmann states specifically that the differences between the higher and lower strata are far from restricted to the broad distinctions between inorganic, organic, conscious and spiritual. 'The higher elements that make up the world', he wrote, 'are stratified in a similar way to the world itself.' This means, for our purposes, that each step in phylogeny that leads from a creature of a lower order to one of a higher order is basically the same in nature as the coming into existence of life itself.

3.3 *Violations of the rules of categorical as well as of systemic analysis*

The agreement between the ontology of Hartmann and phylogenetic investigation emerges most strikingly when one compares their methodological rules, and at the same time the many ways in which these rules are being broken. The ontologist is concerned to describe external reality in terms which are adequate to the phenomena and do not ascribe to things attributes that they do not possess, nor ignore attributes that characterize them. In Hartmann's words: 'It is patently obvious to any unprejudiced observer that the world is built up by stratifications. This was realized early on. The only reason why it was not accepted without opposition was because it conflicted with the premise of speculative philosophy that the world is a unity.'

It is, for instance, an act of metaphysical speculation when radical mechanicists attempt to explain everything that happens in this world on the basis of the categories and laws prevailing in classical mechanics which simply do not suffice for this undertaking. If the mechanicist neglects the categories and laws which are exclusive properties of the higher strata, or even denies their very existence, he commits the well-known but apparently ineradicable fallacy of an 'upward' transgression of the limits imposed upon us by the 'stratification' of the real world. All so-called '-isms', such as biologism, psychologism, etc., presume to explain the laws and processes proper to higher levels in terms of categories derived from lower levels, which is simply not possible.

Likewise, a distortion of the phenomenal world results from transgressing the boundaries between levels of existence in the opposite direction. This reciprocal error is described by Hartmann as follows: 'The base of the entire world picture is then chosen on the level of conscious existence — on the level on which man experiences his own subjective life — and from there the principle is extended "downwards" to the lower level of reality.' All pan-psychistic world views, like Leibniz's system of monads, Jakob von Uexküll's *Umwelt und Innenwelt der Tiere* (1909) and even Weidel's ingenious attempt to solve the problem of body and soul, make this mistake of trying to explain the diversity of the world on the basis of ontological or phenomenal principles of one single kind.

The urge to do this, and thus to evolve a unitary world image, is obviously irresistible to many thinkers. It is impossible to explain in any other way how a man endowed with common sense could conceive the idea of denying that a dog or a chimpanzee possessed the faculty of subjective experience, as Descartes did, or of attributing such a faculty to iron atoms, like Weidel.

All modern systematic phylogenetic research into the formation of new system characteristics and the unilateral relationship between different levels of integration makes it clear that causal analysis appropriate to the investigation of a living system produces results, and is bound to a methodology, which are closely related to those of Hartmann's categorical analysis of the 'strata of existence'. Indeed, one might even claim that it is analysis of this kind that makes us understand why overstepping the line separating one level from another leads us so seriously astray. It becomes quite obvious why it is impossible to deduce the characteristics of a higher system from those of a lower system, and also why it is pointless to delve into subsidiary systems or investigate the primitive ancestors of advanced organisms in search of characteristics and functions which only came into existence at the creative moment of higher integration. And it is naturally even more pointless to postulate the existence of such characteristics and functions on a lower level.

One typical error of this kind will concern me particularly in the course of this book — namely, the persistent attempts of certain psychologists and behavioural scientists to prove the existence of adaptive learning, not just in primitive creatures that are incapable of it, but also in systems of more advanced organisms which are not only incapable of being modified by learning but whose phylogenetic programme makes them resistant to all modification. A psychologist not versed in biology, owing most of his practical knowledge about animals to his experience of human beings and the higher mammals, and brought up in the orthodoxy that reflex and conditioned reflex are the most elementary factors in all animal and human behaviour, may well regard it as a matter of course to ascribe at least some traces of elementary conditioned responses even to protozoa and lower invertebrates, and he may cling to this illusion with the fervour born of a need to believe in a unified world picture. The strength of this urge also explains the large number of sometimes disastrous self-deceptions that have found their way into attempts to prove that even the lowest of living creatures can 'learn'.

3.4 *The fallacy of thinking in opposites*

I showed in the previous section that the various levels of existence have varying characteristics according to the different levels of integration of the systems in question. I also discussed the errors that arise when, seeking a unitary explanatory principle behind the world, we try to explain lower and more primitive systems on the basis of principles applicable only to higher systems, and vice versa.

We must now turn our attention to the opposite error — namely, that of overlooking what is common to all levels of existence. To think in antithetical terms, opposing alpha and non-alpha, is apparently as ingrained in the human mind as is the urge to seek unifying principles, to which it acts as a kind of counterbalance.

The unilateral penetration of the different levels of existence from lower to higher allows us to make two kinds of statement about what is common and what makes for differentiation. One can say, for instance, that all life processes are chemical and physical phenomena; that all the subjective processes of our experience are organic and physiological, and therefore also chemical and physical; and that the whole of man's emotional and intellectual life is thus ultimately an event on all these basic levels. It would be equally correct to say that 'in essence' — that is, in respect of the ontological and phenomenal principles which are quite particularly their own, and which set them above all other chemico-physical phenomena — life processes are something entirely different; that nervous reactions arising from personal experience are utterly different from inanimate, physiological nervous processes, or that man, as a creature endowed with a cumulative tradition and supra-individual knowledge, is fundamentally different from his nearest zoological relations.

The seeming contradiction between these two sets of perfectly true statements indicates the existence of a pseudo-problem, which can prove a serious hindrance to the advance of human understanding.

Its resolution is one of the most important results at which, independently of each other, the ontology of Nicolai Hartmann and the causal-analytical investigation of living systems have arrived: namely, that the law of contradictory opposites does not apply to the unilateral penetration of strata of levels of integration. B is never non-A but always $A + B$; C is $A + B + C$, etc. Although it is inappropriate to use mutually exclusive concepts when dealing with different levels of existence, numerous such pairs of terms have crept into our thinking and are to be found in both our scientific and our everyday language — mind and nature, body and soul, man and animal, etc.

It will be recalled that in the illustration of the electric circuits (p.31) the two systems (*a*) and (*b*) were coupled together in a single system; the new system consisted of the component parts of the other two systems but had characteristics of which there was not the slightest trace in either of these two systems. This should be as easy to understand as the fact that similar *fulgurationes* have constantly taken place in the course of phylogeny, and that although new system characteristics appear, the characteristics of the old system continue to survive.

There are philosophical anthropologists who are clearly unable to grasp this, and who indulge in endless, fruitless discussions of whether man differs from animals in essence or only 'in degree'. They do not see that every new system characteristic, like the oscillations in our example, always signifies a change in essence, not in degree. A warm-blooded creature with its constant blood temperature is essentially different from its poikilothermic ('variable temperatured') ancestors; a bird's wing is essentially different from the reptilian arm from which it evolved; and precisely in this sense, and no other, is man essentially different from other anthropoids. Whenever he heard anyone use the words 'man' and 'animal' as opposites, Heinroth used to interrupt the speaker politely and ask: 'Forgive me, but when you talk of the animal, do you mean an amoeba or a chimpanzee?'

3.5 *Summary of the preceding two chapters*

Summarizing the last two chapters, the one on the creation of new system characteristics, the other on Nicolai Hartmann's philosophy of 'strata of existence', I must emphasize three facts which are relevant to the following study of the structure of phylogeny of cognitive mechanisms. Like all processes in life, those of acquiring and storing relevant information take place on many different levels and are interlinked at many points. In our investigation of these processes we shall frequently find ourselves confronted with the following three facts.

First: even the simplest of systems are capable of working independently, just as the simplest of organisms are, and always have been, capable of survival; otherwise higher organisms could never have evolved from them.

Secondly: a new and complex function often, though not always, arises through the integration of a number of existing, simpler functions which were already independently visible, and which, far from disappearing or losing their importance, continue to function as essential components in the new entity.

Thirdly: it is idle to look in these original systems or in primitive organisms for those system characteristics which come into existence only at a higher level of integration.

Chapter four
Short-term information gain

4.1 *The functional limitations of the genome*

In spite of its almost unlimited capacity, the trial-and-success method of the genome would not in itself be sufficient to maintain living systems in a continuous state of adaptedness that would guarantee their survival. The cognitive mechanism of the genome is unable to cope with any rapid change in the environment, because it cannot 'know' anything about the result of any of its experiments until the passage of at least a generation. Thus it can only produce adaptations to environmental parameters which will remain reasonably constant over a period of time: the generation represents, so to speak, the 'dead time' before the cognitive mechanism of the genome begins to respond to an external influence.

As was suggested in the Prolegomena and discussed in more detail in Chapter 1 (see p.25), there are a great many well-adapted mechanisms which acquire and exploit information but do not store it. The particular nature of these mechanisms is often overlooked, because the functional analogies between the simplest, most basic form of acquiring knowledge, i.e. that of the genome, and the most advanced cultural forms of human cognitive activity, cause us to forget all too readily that between these two qualitatively different processes there is a level of essential cognitive functions, which conveys information about instantaneously arising conditions in the environment, thereby providing the basis for all higher processes of experience and learning. These processes, which occur in all living organisms, including bacteria and plants, form the substance of this chapter.

4.2 *The regulating cycle or homeostasis*

All physiological mechanisms which gather short-term information, and thus bridge the 'dead time' in the genome, are enabled to perform this function by virtue of structures which owe their nature to the trial-and-success method of the genome. Here there is a problem. For since even the simplest form of acquiring instantaneous information — namely, that of the cycle, or homeostasis — depends on the structure of the genome, i.e. on the results of its trial-and-success method, and since at the same time life itself is virtually inconceivable without any regulating cycle, we find ourselves facing a problem which differs from that of the chicken and the egg only in so far as it raises a real and relevant question.

Reference was made above (p.30) to the regulating cycle, or cycle with negative feedback. In the world of organisms there are innumerable such feedback loops in all conceivable degrees of complexity. These extend from simple mechanisms that maintain a state of equilibrium on a chemical level, to sophisticated organizations in which the most complex processes of sense organs and central nervous system ensure, through the behaviour of individuals or whole societies, the maintenance of a *Sollwert* (reference value) or 'equilibrium position', such as the optimum density of population for the survival of a particular species. V. C. Wynne-Edwards has demonstrated this in various contexts.

Whenever an organism regains its equilibrium after it has been disturbed, or maintains it in the face of disturbance, it has received and utilized information about the nature and extent of the change in its environment. When, for example, an animal increases its breathing rate in a rarefied atmosphere, or is offered too much food and refuses to eat, this means that it is informed not only about its own needs but also about the external 'market situation' in these 'commodities' at the time.

Like many other mechanisms through which an organism receives information about current conditions in its environment, the cycle can repeat itself any number of times without changes to its function. In other words, leaving aside signs of age and deterioration, one finds that the structure programmed by the genome remains permanently and precisely the same. The information which the apparatus is built to acquire and exploit is immediately utilized but not stored.

4.3 *Irritability*

Except for certain forms of homeostasis, all processes by means of which an organism acquires and exploits instantaneous information depend on its capacity to respond to so-called stimuli. Stimulus and stimulation have been defined in many different ways. In general, however, we take them to refer to an external influence which triggers off a process that does not rely on the 'stimulus' as a source of energy and which can consist of movements (or a change in any existing motion) or in secretion. One usually associates irritability with the elicitation of movement, at least in lower organisms. It is only the division of labour between muscles and nervous system that enables advanced animals to receive and exploit stimuli without immediately responding with movement.

It seems to be unknown whether in the most primitive organisms motility ever exists without irritability, and bacteriologists have not been able to answer my question as to whether there are micro-organisms that move yet are incapable of responding to stimulation. Motility, particularly locomotion, could conceivably increase the organism's chances of gaining energy, even without any accompanying acquisition of information.

In most living organisms, however, the capacity to respond to stimuli is closely linked with that of locomotion. The primary and most important function of locomotion is to enable the animal to escape from danger. Possibly an even more primitive function of physical movement is when by maximal contraction of the body the organism exposes as small a surface as possible to the threatening external forces and turns the folds in the surface into a thick protective skin. This type of avoidance response, common among static or slow-moving organisms, is frequently accompanied by secretions which help to protect the surface. Some protozoa, coelenterates and other invertebrates, such as snails, behave in this way.

Irritability is the basic prerequisite of all those processes that acquire and exploit instantaneous information without storing it, and also of all those that occur in the central nervous system and form the basis of the highest functions of learning and memory. These will be discussed later.

4.4 *Amoeboid response*

Strangely enough, the simplest and most primitive stimulus-elicited motion that we know of in the organic world is orientated in all three dimensions of space. The amoeba, consisting merely of a naked mass of protoplasm, moves by reducing the thickness of its outer layer, the ectoplasm, at one particular point; a blunt outgrowth is then formed which, as the ectoplasm becomes still thinner, develops into a so-called pseudopod ('false foot'). As the resistance becomes less, the content of the cell moves into the pseudopod, the base of which extends, and the whole cell thus moves slowly forwards. Corresponding to the outgrowth of the pseudopods, a contraction and thickening of the ectoplasm takes place at the posterior end of the amoeba.

This formation of pseudopods and the corresponding motion of the amoeba used to be explained by assuming that the principal cause was changes in the surface tension. And indeed, one can make models to imitate the whole process very accurately by using globular drops and varying their surface tension. After observing amoebae in relatively natural conditions for some time, however, I concluded that this was too simple an explanation. On the basis of my direct observations I maintained that the plasma of the amoeba constantly changed from a condition of sol to one of gel and vice versa; within the cell it flows out into the pseudopod, then, in the way that flowing lava hardens, it congeals when it makes contact with water, soil or some other feature of its environment. What superficially appears as a decrease in surface tension, and, indeed, functions as such from the mechanical point of view, is the partial liquefaction of the gelled ectoplasm, which begins internally at the point where a pseudopod is to be formed. This view, based on a series of simple observations, has since been corroborated by the work of L. V. Heilbrunn. When a noxious stimulus touches the surface of the organism, causing it to contract and start crawling away, this is achieved by the gelling of the liquid plasma immediately adjacent to the point of stimulation. The contraction is caused by the fact that the gelling is accompanied by a slight decrease in the volume of the protoplasm, which has a mechanical effect equivalent to that of an increase in surface tension.

The flowing of the plasma is, however, not due only to the changes in pressure caused by such processes, for it is also found in the cells of plants which are encapsulated in a rigid cellulose sac and in which the pressure is uniform.

If one observes an amoeba in its natural habitat, not on a slide but moving freely in a petri dish, one is amazed at the versatility of its behaviour and its adaptability. If it were the size of a dog, said H. S. Jennings, the greatest expert on protozoa, one would not hesitate to attribute to it the power of subjective experience. Still it only has this one, single manner of movement by which to deal with all environmental situations. One must keep in mind that it is one and the same mechanism by means of which the amoeba avoids damage in 'fear' of being injured, moves towards a source of positive stimulation, or, in the optimal situation, embraces an object that emits positive stimuli and greedily consumes it. It is the same mechanism that is used by the amoeba both to escape and to eat.

The adaptive information that underlies the amoeba's apparent intelligence is based entirely on its ability to respond selectively to different stimuli, and the extent of its response can vary so greatly that an observer may fail to see that the mechanism is the same. An amoeba which, directed by the influence of a change of temperature, acid content, salinity or something similar, crawls sluggishly towards a more favourable habitat, makes a quite different impression from one which 'pounces' on its prey or is 'cunningly' preparing to trap an agile ciliate in its pseudopods. This great variability in behaviour and effortless power of three-dimensional orientation make the amoeba seem a remarkably 'intelligent' creature, yet these are, so to speak, not 'individual achievements' on the animal's part, for they derive from faculties that belong to protoplasm as such, and thus also to any creature that consists entirely of protoplasm.

Only when the amoeba's ability to develop a functional front-end and rear-end at any part of its body is sacrificed in the interests of a rigid structure, in particular the elongated, streamlined form of all creatures that move quickly in water, does the fresh problem arise of how such a structure can be effectively steered in all three dimensions of space. Among multicellular organisms there are only a few radially structured creatures, like the starfish, which can move in any direction they wish, albeit only two-dimensionally. If that out-of-this-world creature, the octopus, appears at first sight to be free of the restrictions imposed by rigid structures on all higher organisms, and if, like the amoeba, it really is free to move three-dimensionally in any direction it chooses, then it owes this faculty, not to the absence of structures but to their profusion and its complete control over them.

4.5 *Kinesis*

We must now discuss a series of orientating mechanisms which enable animals with fixed front and rear ends to use their faculty of locomotion to seek out localities in which a gain of energy is probable, its loss improbable. It may seem remarkable that this can be achieved without any change in the direction of the movement, yet it is possible. An organism moving about at random, which accelerates when conditions in the environment become adverse and decelerates when they are to its advantage, achieves the desired effect simply by means of a change of speed. The undesirable effect that a stretch of very bad road has on the speed and hence on the density of motor traffic may serve as an illustration. If the cars were paramecia in the vicinity of decaying vegetation this behaviour would develop distinct survival value. This simple means of promoting the longest possible stay in the most favourable possible environment was called by Fraenkel and Gunn 'kinesis' (i.e. movement).

In many organisms this process is made the more effective by the fact that when animals which do not move in a straight line but in a kind of zig-zag pattern enter a more favourable environment, the angle of their more-or-less random changes of direction increases, and they consequently explore the new environment more thoroughly. This process, known as klino-kinesis and usually combined with simple kinesis, is found in many different creatures — not only in protozoa but also, in a highly developed form, in various woodlice, which belong to the higher crustacea. Formally and functionally similar modes of behaviour, though based on far more complex sensory and nervous processes, are also to be found in mammals — think, for example, of grazing ruminants, or of people looking for mushrooms.

The mechanism of kinesis is of remarkable simplicity. All that is needed is for a single receptor to have a purely quantitative effect on a single type of movement. This, as far as I can see, is the simplest process by which a non-amoeboid organism able to move in all directions can acquire and exploit instantaneous orientation information. What the organism learns about its environment can be expressed in the simple phrases, 'It's better here' or 'It's not so good here'. The consequences that it draws from this 'knowledge' are equally simple: 'Let's stay here a while', or 'Let's get away from here'. However, it learns nothing about the direction of the gradient that causes the improvement or the deterioration of its environment.

4.6 *Phobic response*

There is a group of lower animals which react with stereotyped turning response whenever their locomotion takes them along a stimulus gradient signifying a rapid deterioration of environmental conditions. In this case the organism does learn something about the direction in which the thing-to-be-avoided lies. But when, on the other hand, it moves along a gradient signifying an improvement in its environment, it makes no response, unless, as is often the case with protozoa, kinesis is also present. Only when the 'stupid' creature has emerged from the favourable area into a less desirable one does it react with its avoidance response. Otto Koehler wittily likens it to a man who accepts a raise in salary without a word but sends up howls of lamentation if his salary is cut.

As Jennings showed, and as Kühn emphasized in his classic work *Die Orientierung der Tiere im Raum*, the extent of the turning response is not controlled by the direction from which the impinging stimulus comes. The paramecium, for instance, behaves as follows: first the beat of the cilia covering the body of the creature is reversed, causing it to swim back for a while along the path it has just come; then the cilia on one side of the body, together with those in the area surrounding the oral groove, begin to propel it forwards again. This means that, to start with, the paramecium moves neither forwards nor backwards but remains circling in the same spot, describing a spiral movement. After a while, depending on the strength, not the direction, of the stimulus, the cilia revert to their original beat, and the animal swims off in the direction in which its longitudinal axis happens to be pointing at the moment.

The spiral movement may happen to have covered 360 degrees by the time the animal resumes its forward motion, in which case it will swim back into the source of the stimulus. It may also happen that the new direction is even less favourable than the former and leads to an even steeper gradient of adverse stimuli. In both cases the animal repeats its response. This behaviour, called by Kühn a 'phobic' response, is far superior to kinesis — though certainly not to the amoeboid response — as far as the amount of information which it conveys to the animal is concerned. It tells not only that a particular environment is unfavourable but also in which direction conditions are even less favourable; it does not indicate, however, the direction of the least favourable conditions, still less where a favourable environment is to be found. As the phobic response not only influences speed of movement, like kinesis, but also direction, it can keep the creature permanently away from an unfavourable zone or permanently within a favourable one, and not merely, like kinesis, reduce the time it spends in the one or extend the time it spends in the other.

The information about what is a favourable and what an unfavourable zone comes to the organism, in the phobic response as in the amoeboid response and in kinesis, via that part of the process which absorbs and so to speak, filters the stimuli. The problems concerned with this function will be discussed in subdivision 8 of this chapter.*

4.7 *Topic response or taxis*

On a far higher plane, as concerns both the quantity of instantaneous information acquired and the complexity of the processes involved, is a type of orientation response known, after

---

*This should not give the impression that the phobic response is the only orientation mechanism that the paramecium has. As Waltraud Rose has shown, it is fully capable of changing course by an appropriate degree without demonstrating a phobic response, provided that its movement carries it into unfavourable conditions at a sufficiently acute angle and that the gradient itself is not too steep. The phobic response is only rarely observed in natural conditions. As Otto Koehler has demonstrated, the paramecium is also able to distinguish between stimuli received at its front end and those received at its rear end; it reacts phobically only to the former, while to the latter it reacts with a perfectly logical acceleration in a forward direction. When the stimuli are exceptionally strong, it can no longer make this distinction: if one suddenly applies a very hot pin to its rear end, it will jump backwards just as if one had applied the pin to its front end. This 'shock reaction', as Koehler called it, can cause the animal's destruction, though only in circumstances that hardly ever occur under natural conditions.

Alfred Kühn, as topic response or taxis. Many important investigations into the spatial orientation of animals have been carried out since the publication of Kühn's classic work. Especially significant for understanding the underlying physiological processes are those of Mittelstaedt and Jander, who approached the subject from the point of view of cybernetics. To summarize the results of their work, however, lies beyond the scope of this book.

The simplest topic response, called by Kühn 'tropo-taxis', derives from the process whereby the organism continues turning until equal stimulation is achieved from two symmetrically placed receptors. A flatworm which displays a positive tropo-taxis to currents which carry the scent of food, turns until the current strikes both sides of the tip of its head with equal force and then crawls upstream. If one reproduces this situation artificially by directing two jets of water symmetrically at the worm's head through a bifurcated tube, the animal will crawl along between the two jets, thus missing both sources. From this simple mechanism, which almost corresponds to Jacques Loeb's theory of tropisms, numerous mechanisms of increasing complication lead up to sensory nervous systems with highly complex feedback mechanisms. A good example, as Mittelstaedt showed in years of research, is the apparatus by means of which the praying mantis strikes at its prey.

Common to all these taxes, the simplest and most complex alike, is the fact that the creature turns directly, without any process of trial and error, in the direction most favourable for its survival. In other words, the size of the angle through which it turns is directly dependent on that formed by the creature's longitudinal axis and the direction of the impinging stimulus. This 'angle-controlled' turning is characteristic of all taxes.

Whereas the phobic response gives the organism only information about the direction in which it should *not* go, and none about the innumerable other directions from which it could choose, the topic response informs it directly which of all these possible directions is the best. In terms of the amount of information acquired, therefore, the taxis is far superior to both the phobic response and to kinesis, but not — to emphasize this once more — to the response of the pseudopods of the amoeba.

4.8 *The innate releasing mechanism*

In the above section on amoeboid movement, kinesis and phobic response, I had cause to refer not only to the information-gaining function of the respective locomotor process but also to the function of the physiological mechanism that releases it. The organism requires not only those nervous structures which underlie its teleonomic motor pattern but also an apparatus to receive those stimuli which will tell it at what moment and in what circumstances the relevant behaviour pattern is likely to ensure survival.

When, in the physiology of the nervous system, the important principle of the reflex arc was discovered, it seemed an obvious step to put all processes that elicit movement into the category of reflexes, and when Pavlov expounded the no less important principle of conditioned responses, it seemed natural to interpret all relevant and apparently innate responses, i.e. those not produced by a learning process, as 'unconditioned reflexes'. This is not entirely incorrect but it obscures the real problem. In an animal with a centralized nervous system it is very likely that the receptor, the apparatus that receives the stimulus, and the effector, which produces a relevant motor response, are linked by a nervous pathway to form a system that corresponds very well to the general idea of a reflex arc. It has been discovered in many cases exactly how such a pathway runs and how many elements it consists of.

Our problem, however, does not lie in the reflex process itself but, in a manner of speaking, before it even begins, in the receptor. How does the organism 'know' which response has to follow which stimulus in order to fulfil its survival function? How, for example, does it happen that the amoeba does not devour all tiny objects but, with rare exceptions, only those which serve as food? How does a micro-organism that survives through kinesis know when and where to swim quickly and when and where to swim slowly?

We have to assume that each such motor response has a mechanism that filters the stimuli, i.e. allows only those to impinge which are reasonably likely to define the particular environment in which the elicited behaviour pattern can have its proper effect. This mechanism might also be compared to a lock that can only be opened by a special key — hence the term 'key stimulus'. The physiological apparatus that filters the stimuli is known as the 'innate releasing mechanism'.

In the case of unicellular and elementary multicellular organisms with a comparatively restricted stock of behaviour patterns mainly concerned with looking for food and sexual partners and with avoiding dangerous situations, the powers of selectivity of the innate releasing mechanism are not greatly tested. Nevertheless, an amoeba responds selectively to a whole number of different stimulus situations, albeit with only quantitatively different motor patterns. Infusoria, on the other hand, including the paramecium, are far less flexible. By means of its phobic and topic responses this creature seeks an environment containing *inter alia* a particular concentration of H-ions. The commonest acid found in nature is carbonic acid, the highest concentration of which is found in waters in which paramecia flourish, especially in the vicinity of rotting vegetable matter, because the bacteria that live on this matter give off carbon dioxide. This relationship is so dependable, and the occurrence of other acids, let alone toxic ones, is so rare, that the paramecium manages admirably with one single item of information, which put into words would say that a certain acid concentration signifies the presence of a mass of bacteria on which to feed. The programme of the species cannot, of course, provide for the eventuality that an experimental physiologist may one day add a drop of poisonous oxalic acid to the paramecium's environment.

In higher animals with well-developed sense organs and central nervous system, and with a wide choice of behaviour patterns, more exacting demands are made on the powers of selectivity of the innate releasing mechanisms, particularly if a variety of combinations of stimuli issuing from one single sense organ must elicit different responses. No such problem exists where, as with the female cricket, a sense organ is only capable of receiving one kind of stimulus, which elicits only one behavioural response. As J. Regen has shown, the female cricket hears nothing but the mating call of the male. The young of most species of cichlid fish, on the other hand, as E. and P. Kuenzer have observed, react optically both to the sight of the mother which they follow and to that of a predator of the same size which they have to escape from. Applied to the wrong object, either of these two different behaviour patterns would spell disaster.

Theoretically the female cricket's auditory organ could be linked directly with the motor apparatus; in the case of the fish there must be a filter between the receptor and the effector to keep the two kinds of key stimulus apart. This filter can only be located in the nervous system itself, i.e. between the receptor and effector organs.

We know little about the physiology of this filter, though the researches of Lettwin and his collaborators, together with Eckhard Butenandt's work on the retina of the frog, have thrown considerable light on how the function could have arisen. Recently, Schwartzkopff and his pupils have shown from grasshoppers that the chain of ganglia through which the key stimulus passes do actually operate as a filter.

When one sees in natural conditions how reliably and efficiently an innate releasing mechanism informs the organism which behaviour pattern is in the best interests of its survival in a given set of circumstances, one is inclined to overestimate the amount of information contained in the mechanism. When one observes how 'cleverly' paramecia stay close to the mass of bacteria, or how quickly a newly hatched turkey chick seeks cover at the sight of a bird of prey in flight, or how, when a young kestrel first encounters water, it bathes in it and then cleans its feathers, as though it had done this a thousand times already, then it is almost a disappointment to discover that the paramecium is governed by the acid content, the turkey chick cowers in exactly the same way when a large fly crawls across a white ceiling, and a polished marble surface produces the same effect on the kestrel as water.

The innate information in the releasing mechanism is coded in the simplest possible way without incurring too much danger of the behaviour pattern being released in any other than the biologically adequate situation. The classic example of simple information which is perfectly adequate for the purposes of the animal in question is the innate releasing mechanism that causes the common tick (*Ixodes rhicinus*) to bite. Uexküll has shown that the tick bites everything that has a temperature of 37°C and smells of butyric acid. This characterization of the tick's natural host, i.e. the mammal, is as simple as the prospect is unlikely that the response can be elicited by anything else in the forest.

4.9 *Species-specific instinctive behaviour patterns*

Innate releasing mechanisms play a special role in cases where they elicit a fixed motor pattern. In organisms such as arthropods and vertebrates, whose hard segmented skeleton permits only limited movement, there are always complete innate motor coordinations already programmed in the genome. These are known as fixed motor patterns. Physiologically they are distinguished by the fact that their fixed pattern of movements, contrary to what one would expect, is not activated by a succession of reflexes but by processes which take place in the central nervous system itself, without receptors being involved. Erich von Holst, Paul Weiss and others have made detailed studies of this subject, and recently E. Taub and A. J. Berman have shown that even in primates many motor coordinations, often of a highly differentiated nature, function quite independently of any control by outer or inner receptors. The incidence of afferent control is only of general space-orientating influence in fixed motor patterns, and insignificant so far as the sequence of movements of which the pattern consists is concerned. Even without the experiment of de-afferentation, e.g. severing all sensitive nerves by vivisection, one can deduce this from the frequent phenomenon of 'vacuum activity', in which a motor pattern runs its full course without the normally eliciting object being present. For instance, when building its nest, the weaver bird (*Quelia*) can execute the highly complex movement of tying a blade of grass to a twig without any blade of grass or similar object being present. It is as though the bird had a 'hallucination' of the presence of such an object.

The performance of a motor pattern is not in itself a cognitive process. The pre-adapted motor skill contained in it is there for the animal to use like a well-made tool; the more specialized the purposes for which it is needed, the more restricted its uses become. There are motor patterns of general applicability, such as locomotion, gnawing, pecking or scratching, and on the other hand there are those which are highly specialized to perform one particular function, such as the case of the weaver bird tying a knot, or many of those connected with courtship and mating.

It is particularly in these fixed motor patterns linked to specific functions that the rigidity of their adaptedness, their complete independence of any kind of learning, emerges most clearly. Even an experienced ethologist is time and again amazed when he watches how a hand-reared young animal, which he knows for certain cannot have acquired any kind of relevant experience of its own, performs a complete pattern with absolute assurance and efficiency. Heinroth described how a hand-reared, newly fledged goshawk that had just learned to fly pounced on a pheasant that was on the point of flying from the table to the window-sill, killed it in the air and landed with its prey on the corner of a cupboard before its keeper could intervene. Heinroth added: 'This, the hawk's first "official" action, so to speak, made an indelible impression on our minds.' This combination of motor skill and precise 'knowledge' about the situation in which it is to be applied presupposes an immense store of innate information.

The sequence of events I have just described, consisting of the stimulation of an innate releasing mechanism and the subsequent elicitation of a fixed motor pattern, is a functional unit extremely common in the animal world. Heinroth called it the 'species-specific drive action' ('*arteigene Triebhandlung*'). This proved to be an extremely valuable concept, although further analysis of it, together with the discovery that the two components could also be integrated in other kinds of unit, did not come until much later.

In higher animals the species-specific drive action represents the prototype of a cognitive process which, as described in Chapter 1 (p.25), is not an adaptation but the function of an already adapted mechanism. Ready-made information on what is the biologically 'correct' situation is already provided, as is the information which the organism needs in order to deal with this situation. The process that supplies instantaneous information merely says to the animal: '*Hic Rhodus, hic salta*', meaning 'now is the moment' to carry out the particular behaviour pattern.

In higher animals the instinctive behaviour pattern is the paradigm of a linear chain of events which is already functional in a simple form, but which through integration with other, equally simple processes has led to the creation of truly epoch-making new functions. One only finds the simple chain in its pure form in creatures which, like jumping spiders (*Salticidae*) and many insects, carry out that particular pattern only once in their lifetime.

In higher animals there is a further element to be considered, namely, the search for the elicitatory stimulus situation which Wallace Craig called 'appetitive behaviour'. It is conceivable that a behavioural chain with this added initial component could also serve a survival purpose in purely linear sequence, but I know of no concrete example. Wherever appetitive behaviour occurs, one also almost invariably finds that the outcome of the action has feedback effect on the behaviour that preceded it. This leaves us, in fact, with the cyclical pattern on which learning proper, i.e. learning by trial and error, is based (see Chapter 6).

4.10 *Further systems based on innate releasing*
    mechanisms and fixed motor patterns

As I have said, Heinroth and I have long considered the species-specific drive action to be the most basic and most important factor in all instinctive behaviour, both of animals and of human beings. The discovery that it comprises two physiologically different components followed from Holst's discovery that the fixed motor pattern does not consist of chains of unconditioned reflexes, as had been hitherto assumed. Holst demonstrated not only that the motor coordination assumes precise form without the help of reflexes but also that it needs no external stimuli in order to be set in motion. Tenches whose sensory spinal nerves have been severed make normal swimming motions; the nervous system of an earthworm, separated from the rest of its body and suspended in Ringer's liquid, continues to send out in regular sequence the impulses that would have stimulated the worm's muscles, had they still been present, to perform fully coordinated crawling motions. The behaviour pattern is thus produced by generation and coordination of stimuli within the central nervous system itself. What Holst called the 'cloak of reflexes' only serves to adapt the spontaneous behaviour patterns welling from within the CNS to the spatial and temporal conditions of its environment.

A motor pattern forms a fixed framework whose structure contains nothing but phylogenetically acquired information. It is only made functional by the many mechanisms for acquiring instantaneous information, which release it in the appropriate circumstances and regulate it in time and space. I regard this discovery as the birth of ethology, since it defined the Archimedean point from which our analytical investigations started.

The physiological discoveries outlined above cast new light on the processes which combine with the fixed motor pattern to form functional units. As long as instinctive behaviour patterns were held to be chains of unconditioned reflexes, the act of release was seen as the first link in the chain, one reflex among many. Physiologically it did not appear to be very different from the subsequent ones and hence not remarkable. But when it was realized that the endogenous stimulus-production of each such behaviour pattern goes on uninterruptedly and has to be permanently inhibited by specific higher centres — that is to say, when it became clear that the release of the motor pattern is basically merely the de-inhibiting of its spontaneity, the question arose, what is the physiological mechanism that causes this de-inhibition?

In many lower animals the most important function of the highest centres of the nervous system is to exercise a permanent inhibiting effect on the various endogenous-automatic behaviour patterns of the organism, and on the basis of instantaneous external information to 'set the pattern off' at the appropriate moment. An earthworm deprived of its 'brain', i.e. its supra-oesophageal ganglion, continues to crawl and is unable to stop; a crab operated upon in the same way cannot stop eating as long as food is available, and so on.

The discovery of the endogenous stimulus-production of centrally coordinated motor patterns demanded a reassessment not only of the process of their de-inhibiting but also of a large number of other highly significant phenomena. From my own observations, as well as from those of Heinroth and Lissmann, it had long been known that if a fixed motor pattern falls into disuse for some time, the threshold of its releasing stimuli does not remain constant but gradually falls. The pattern becomes easier and easier to release and begins to respond to substitute stimuli; in extreme cases it even starts without any demonstrable stimulus at all. At the time I was studying these phenomena I still believed in the chain-reflex nature of fixed motor patterns, though I did pay particular attention to phenomena which could not be accommodated by the chain-reflex theory. It is these phenomena that can be convincingly explained with the help of Holst's discoveries. The lowering of the threshold of the releasing stimuli is, however, not the only effect of withdrawing from the animal for a time the condition adequate for a particular behaviour pattern. It excites the organism as a whole, causing it to search actively for the key stimuli. This is the 'appetitive behaviour' we discussed at the end of the previous section.

Thus what 'drives' the animal — the 'drive' in Heinroth's term ('drive action') — is in many cases far from identical with the powerful general instincts in animals and humans, such as hunger, thirst and the sexual urge, whose physiological causes are comparatively easily traced to the absence of vital substances or to the 'detumescence instinct' set off by internal pressure within hollow organs like the bladder, etc. Rather, each individual instinctive behaviour pattern, however small, represents an autonomous motivation element if — to use an awkward phrase coined by Freud's translator — 'it is not being abreacted'. This plays an important part in the development of conditioned responses, as will be seen from Chapter 6.

Tinbergen and Baerends were the first to show that the three important components of an instinctive behaviour pattern — appetitive behaviour, the stimulation of an innate releasing mechanism and the consummatory performance of the motor pattern — can be linked in another and more complex manner. A long sequence of behaviour patterns is often programmed in such a way that the initial appetitive element leads to a stimulus situation which does not then immediately release the motor pattern but first a further kind of appetitive behaviour. A falcon flies around in search of prey — appetitive behaviour of the first order. Coming upon a flock of starlings, it circles high above them and then swoops at them in a manoeuvre calculated to isolate one particular bird from the rest — appetitive behaviour of the second order. Only when this has been achieved is the falcon at the point where it can move to the next stage — namely, to 'bind to' the prey, to use the falconer's expression, which is then followed by further motor patterns, first of killing, then of plucking it, and finally of eating it. It is important for our conception of the nature of fixed motor patterns to realize that in a sequence of this kind any single pattern — except the last one — is at the same time the end or goal of the appetitive behaviour preceding it as well as the means to reach the next one.

Such a system leads an animal in a programme well organized in phylogeny from one releasing situation to another, whereby — and this is important — the movement is always from a more general, more accessible situation to a more specialized one, which would be difficult to find if the first releasing mechanism, switching from appetitive behaviour of the first order to that of the second order, did not point the way to it. This concept of a 'hierarchically arranged instinct', as Tinbergen called it, is far from being merely a metaphysical construct, for the number of releasing mechanisms involved is experimentally verifiable, and the fixed motor patterns can be precisely described.

The chains of appetitive and releasing mechanisms involved in a system of this kind can also be bifurcated — that is to say, after the stimulus situation sought by the appetitive behaviour of the first order has been achieved, several further behavioural patterns become possible, i.e. several further appetites can be aroused. As the spring days grow longer, the hormones of the stickleback awake its readiness for reproduction; it begins by moving from deep water to shallow, and keeps on moving until it has found a piece of quiet, shallow water with vegetation growing in it. It then becomes 'territorial' — that is, it occupies a particular spot and never leaves it. At the same time a new readiness to perform certain behaviour patterns is awakened, and this manifests itself, both to its fellows and to scientists observing its behaviour, in a change of colour. The stickleback is now ready to build a nest, to fight off rivals, to court a female, take her to the nest and mate. The stimulus situation produced by the appetitive behaviour of the first order thus forms the basis for simultaneous emergence of a number of new appetitive elements. Tinbergen has made detailed studies of the instincts of sticklebacks, while Baerends has investigated those of the sand wasp (*Ammophila*).

An instinct system of this kind is far more flexible than Heinroth's 'species-specific drive action', because each of the successive releasing mechanisms acquires information about conditions as they are at the moment and adapts its behaviour accordingly. The many taxes involved in the process as a whole also contribute considerably to this increase in information.

Hierarchical instinct systems can be adapted to save both time and energy by missing out certain links in the behavioural chain. If, in the example above, our falcon comes upon a single starling, it naturally saves itself the trouble of isolating it from the rest of the flock and attacks it at once. Baerends showed, however, that 'short cuts' of this kind could not always be induced in the sand wasp, which often persisted in its regular sequence of actions.

The often remarkably 'intelligent' adaptability of these systems is, we must once again emphasize, the product of functions concerned with the reception of instantaneous information. The system is adapted, but the openness of its programme admits a wealth of combinations of behaviour patterns relevant to the animal's survival even without any adaptive modifications of its machinery. This places it clearly in the category of cognitive processes that function independently of learning processes — which is the subject of this chapter. This independence can be seen from the many cases in which complex, hierarchically organized behavioural chains are carried out only once in the lifetime of an individual animal, as revealed in the detailed study by G. W. and E. G. Peckham and J. Crane of the mating habits of various spiders.

It is also shown in Ernst Reese's investigation and film of the larvae of hermit crabs, which, swimming on the surface of the sea as glaucothöe postlarvae, perform on their first encounter with a snail shell the whole sophisticated behaviour pattern by which an adult hermit crab discovers a snail shell, examines it, cleans it and finally occupies it.

The essential independence of these instinctive behaviour patterns from learning processes has not prevented them from becoming the very foundation on which learning mechanisms have been built up. Their 'unilateral' relationship to highly developed learning processes is analogous to that between the different levels of integration described earlier (p.33).

The mechanisms for acquiring instantaneous information function independently of the learning processes which later become superimposed on them, but they constitute the prerequisite for the coming-into-existence of these processes, and the adaptive modification of behaviour (habituation, sensitation, etc.) discussed in the following two chapters would be impossible without them, in particular the function of 'conditioning by reinforcement'. The creation of this highest form of learning presupposes the existence of fully operational systems consisting of appetitive behaviour, innate releasing mechanism and the consummatory situation. Without the presence of these three components there could never have been the *fulguratio* of feedback which allows successful consummation to change adaptively into the machinery of preceding behaviour and constitutes the essence of the conditioned response in the narrow sense.

4.11 *Summary of chapter*

Of the cognitive mechanisms discussed in the first three chapters, only the genome, through its method of trial and error, has the power both to acquire and to store information. There is scarcely any limit to the quantity of such information, but the time required to draw from it the conclusions necessary to ensure survival spans at least the life of a generation. Living systems cannot therefore maintain their adaptedness — or at least could only do so in conditions of inconceivable invariability — without being able to receive and exploit, through short-term mechanisms, information about present conditions in the environment.

The oldest, commonest, most basic of these mechanisms is the cycle, which uses negative feedback to keep certain conditions constant within the organism and independent of fluctuating circumstances outside it (i.e. homeostasis).

Mobile organisms are able in various ways to come to terms with the special parameters of their environment, as shown by amoeboid response, kinesis, phobic response and taxis, all four of which depend on irritability.

These processes are followed, on a higher plane, by innate releasing mechanisms, fixed motor patterns and other, more complex systems built up on these two elements.

In contrast to the cognitive functions both of the genome and of higher processes involving learning, none of the mechanisms discussed in this chapter is capable of storing information. Their functions are not processes of adaptation but operations of already adapted structures. They are protected against any kind of modification, necessarily so, because, being present before all experience, they are the foundation of all experience. In this respect they correspond to Kant's definition of what is *a priori*. Similarly, the mechanisms for acquiring instantaneous information function independently of learning processes but are the indispensable foundation for all such processes that develop on a higher level of integration.

Chapter five

Adaptive modifications of behaviour (excluding conditioning by reinforcement)

5.1 *General*

Modification is the name given to any change caused by external circumstances during the individual life of an organism. It determines the external appearance of every individual — the phenotype — on the basis of its hereditary factors — the genotype. It is at work everywhere. It would be hardly an exaggeration to say that every slight difference in the ecological conditions in which two genetically identical individuals develop produces a slight difference in their qualities — i.e. their phenotype. These modifications of the structural blueprint by environmental factors do not, however, necessarily imply changes favourable to the survival of the species in question. The probability that a modification caused *by* a change in the environment represents an adaptation *to* this change is no greater than the probability that a random mutation or new combination of genes will further the chances of survival. If in response to a specific influence an adaptive modification regularly occurs, one can be virtually certain that this specific modifiability is the result of an earlier process of natural selection.

Thus when, for example, our blood has more haemoglobin and red corpuscles at high altitudes, because of the lack of oxygen and the low atmospheric pressure, or when a dog grows a thicker coat in a cold climate, or when a plant growing in dark conditions stretches upwards so as to give its leaves more light, this is by no means due solely to the environmental influences, but also to an in-built genetic programme which has been evolved by the genome through trial and success and is now available to be used in these particular circumstances. It is as though the plant had been told: if the light is inadequate, extend your stem until satisfactory illumination is achieved. Ernst Mayr called this kind of genetic information an 'open programme'.

An open programme is a cognitive mechanism able not only to acquire information about the environment which is not contained in the genome but also to store it. In other words, the ontogenetic realization of the most appropriate option among those offered by the open programme is an adaptive process.

The fact that the open programme acquires and retains information in this way must not lead us to overlook that it requires for this purpose not less, but more genetic information than that required for a closed programme. This can be illustrated by the following analogy. A man wants to erect a prefabricated house without the need to make any adaptive alterations to it — an example of a completely closed programme. The only suitable site is a completely level surface, like the absolutely horizontal lava terraces found on volcanic islands. In such a case the builder needs very few instructions. If, however, the same house has to be erected on an uneven or sloping site, we can imagine what a vast number of additional instructions the builder would require in order to solve the problem, a problem different for every site.

Such an example illustrates the fallacy of thinking in opposites, of treating what is innate and what is learned — 'nature and nurture' — as mutually exclusive concepts. All learning ability is based on open programmes which presuppose the presence, not of less, but of more information in the genome than do so-called innate behaviour patterns. The reason why so many otherwise perceptive thinkers have difficulty in grasping this can only be the general human tendency to think in opposites.

Adaptive modification exists on all levels, from the very lowest organisms upwards. Certain bacteria, for instance, when cultivated in an environment with a low phosphorus content, will promote the growth of those chemical structures in the cell which absorb phosphorus. The bacteria take some time to make this change, but if one suddenly transfers them to an environment rich in phosphorus, they will consume an excessive amount of phosphorus before finally reversing the adaptive modification of the cell structure involved. The cognitive function of this process resembles that of homeostasis to the extent that it provides the organism with information about the 'current market situation'.

This adaptive modifiability of the bacteria makes one think in terms of a learning process, though in general we restrict this term to adaptive modifications that affect behaviour. The acquisition of instantaneous information that cannot be stored, together with all the cognitive processes described in the previous chapter, is not learning. It is essential to all learning processes that an adaptive change should take place in the 'machinery' — that is, in the structures of the sense organs and nervous system that are concerned with behaviour. For it is in this change of structure that the acquisition of information and, since the change is more or less permanent, the storing of that information, resides.

Adaptive modification, particularly of behaviour, is a cognitive process of a special kind, being superior both to the function of the genome and also to all the above processes for acquiring instantaneous information, in that it can not only store information, like the former, but also cope with short-term changes in the environment, like the latter. None of the processes described earlier can do both.

The structures that are changed in adaptive modifications of the behaviour of higher animals are most probably those of the central nervous system. (I shall return later to the improbability of the notion that the results of learning are encoded in chain molecules like the information specified in the genome.) The more complex a living system, the less likely it is that a random change in its structure will have any other effect than that of a malfunction. There is no more complex system known to us in the whole world than that of the central nervous system, which is at the root of the behaviour of all advanced living beings. It is therefore a quite remarkable achievement on the part of evolution to have produced, in this of all systems, so varied an adaptive modifiability. This modifiability derives from a body of incredibly complex structures which are the basis of open programmes and leave the way free for learning processes to operate. There has hardly been a greater error in the history of human thought than the view of the empiricists that prior to every experience the mind is a blank sheet, a *tabula rasa*. Equally disastrous is the apparently opposite, but basically identical error made by many non-biologically-minded psychologists, that learning must form part of every element in animal and human behaviour, however small. Both these errors have the regrettable effect of blurring the central problem in all learning — i.e. how does it happen that it strengthens behaviour which promotes the survival of the species?

5.2 *The testimony of experimental embryology*

An open programme involves a cognitive, i.e. an adaptive function. An external impulse provides the information on the basis of which one of the options in the programme is put into effect, namely the one best suited to the situation.

Experimental embryology has thrown considerable light on this still very mystifying process. A classic example of an open programme with various options is that of the ectoderm of vertebrate embryos. According to their position in the body of the embryo, ectoderm cells can form epidermis, parts of an eye, or a brain together with spinal cord. Every ectoderm cell contains the information necessary to form each of these organs, and which of the programmes is put into effect depends on environmental influence. Left to itself, for example in the case of a piece of cell material cut from the abdominal region of a newt embryo, the ectoderm will only form epidermis; when grafted close to the *chorda dorsalis* — antecedent of the spinal column — it will form a spinal chord and a brain; when later the optic vesicle grows out of the forebrain, the ectoderm forms a lens at precisely the right point. It is easy to prove experimentally that the particular development is 'induced' (Spemann's term) by influences from neighbouring cells: if one grafts a small piece of *chorda dorsalis* under the newt's abdominal skin, a neural tube begins to form in the ectoderm above.

The original range of latent possibilities within a piece of tissue — what Spemann, the first great investigator of these processes, called the 'prospective potency' — is thus greater than its 'prospective significance', for this significance depends on the position in which the pieces of tissue develop. The influence of the position includes one particular line of development from among those available, and once it has been followed for some time, it finally becomes determined — that is, it can no longer be changed, and the prospective potency of the cells is now restricted to their prospective significance.

The various different processes involved in adaptive

modification are all essentially akin to these processes of embryogeny. It is of no great significance whether an inductive influence issues from the vicinity of a piece of tissue within an embryo or from the external environment of an organism. The system always contains genetic information on all the programmes that it is potentially capable of carrying out. The ectoderm is not 'told' by the *chorda dorsalis* how to make a spinal cord and a brain, or 'told' by the optic vesicle what a lens should look like. Spemann's somewhat vitalistic concept of an 'organizer' is thus rather misleading. We now know that inorganic impulses, too, are capable of induction — for example, causing the ectoderm to form one of the abovementioned organs. The same is true of many adaptive modifications, including those affecting behaviour. All learning resembles the process of induction to the extent that certain external impulses lead to the selection from among the various possibilities in an open programme of that which is best suited to the conditions of the environment. And these external impulses are themselves already 'planned for', in the sense that they are built into the programme on the basis of previous adaptive processes.

How firmly and specifically such impulses can be programmed into a particular learning process has been shown in the experiments of J. Garcia and his team, to which we shall return later in another context. Suffice it to say here that one cannot train a rat not to eat a certain food by any form of punishment stimulus, however painful, but one can do so by using stimuli that produce even a slight nausea in its digestive organs.

Learning in its broadest sense, i.e. the adaptive modification of behaviour, is thus similar in essence to the process of induction, but with one fundamental difference: that once the confining process of determination has taken place, the induction can no longer be revoked, whereas behaviour that has been learned can be forgotten, or even reversed by counter-training. Karl Bühler even seriously considered whether this reversibility ought not to be included in any definition of learning. However, there are learning processes which are not reversible but determined once and for all. On the one hand there is the process of imprinting, a form of object fixation; on the other hand there are processes connected with the acquisition of powerful avoidance responses which leave indelible marks as so-called 'psychic traumas', particularly in the young.

Whether or not one calls all teleonomic modifications of behaviour 'learning' is a matter of taste. Some authors even apply the term to the acquisition of knowledge by the genome. In my study *The Innate Bases of Learning* I used it to include all modifications of behaviour that served the survival of the species. As the whole behaviourist school of psychology bases its work on the hypothesis that 'learning by success', or 'conditioning by reinforcement', is the only form of learning, indeed the only process worth examining in the whole field of animal and human behaviour, I consider it appropriate to stress the specificity of this learning process by treating it fully in a separate chapter and restricting myself here to the simpler forms by which individual knowledge is acquired.

5.3 *Facilitation by practice*

We are all familiar with the process of 'running in' an automobile, by which the engine undergoes an adaptive change. A similar process occurs in a number of mechanisms of behaviour. For instance, M. Wells found that a newly hatched cuttlefish (*Sepia officinalis*) caught its prey from the very first moment with complete coordination, but that its response was considerably quicker after it had performed the operation a number of times. Its aim also improved. A similar discovery was made by E. H. Hess from the pecking movement of newly hatched domestic chicks. As Hess demonstrated, the efficiency of the movement is not affected by whether the chick strikes the object or misses it. He fitted the chicks with hoods enclosing prismatic lenses, which displaced the visual field seven degrees to right or left. The animals never learned to correct the displacement but went on pecking in the direction where they expected the object to be, and thus missing it. However, after practice they considerably reduced the scatter of their movements.

5.4 *Sensitation*

On the sensory side, the counterpart of facilitation by practice is that of 'sensitation', the name given to the process by which the repeated elicitation of a response lowers the threshold level of its key stimuli. With its first response the animal is, so to speak, alerted, has its attention aroused. This metaphor also conveys the fact that sensitation is generally shorter term than facilitation by practice.

This state of being alerted will only have a survival value in cases where the releasing stimulus situation can be expected to recur. This applies particularly to escape-releasing stimuli. An earthworm that has just escaped the blackbird's beak will do well to 'keep in mind' that the blackbird is still around. Wells has demonstrated that where the object of the response, be it enemy or prey, is regularly found in large numbers, as is often the case with organisms living in the open sea, sensitation develops a very high survival value. One of the most striking examples of this is what is known as 'feeding frenzy' in deep sea fish such as sharks, mackerel and herring. After swallowing a few victims these fish get frantic and begin to snap savagely in all directions, while the thresholds of the key stimuli fall to such a point that tunny fish, for example, snap at bare hooks. This is the principle on which tunny fishing is carried on in tropical waters.

Sensitation is a widespread form of learning in lower animals, and particularly characteristic, according to Wells, of marine bristleworms (*Polychaeta*), among which there are highly developed predatory forms equipped with efficient sense organs.

With both motor facilitation and sensitation any functional improvement in a system is achieved by the function itself, and this is one of the constituent characteristics of learning. At the same time neither of them possesses a characteristic usually regarded as a constituent element in learning *viz* association. By this is meant the development of a new relationship between two nervous processes which functioned independently before the particular learning process brought them together. Association is a characteristic feature of all the learning processes discussed below.

5.5 *Habituation*

A stimulus situation which on its initial appearance elicits a response of a certain intensity frequently loses some of its effectiveness on its second appearance, and after a series of repetitions its effect may disappear completely. This is known as habituation. The usual alternative term 'sensory adaptation' is misleading and it suggests that the process invariably develops a survival value, which it does not always do.

The waning of the response does not usually depend on whether the key stimulus is followed by a reinforcing stimulus situation or not. In many respects this phenomenon resembles a state of fatigue, and may even have developed phylogenetically from particular symptoms of fatigue. Its high survival value, however, lies in the fact that it actually prevents fatigue of the response from setting in, above all on the motor side.

This is achieved through the fact that habituation affects only a specific type of stimulus. The hydra responds to a set of different stimuli by contracting its body and arms into as small a space as possible: a shaking of the ground, a touch, a slight motion of the water, a change of temperature or a chemical stimulus — all have the same effect. If, however, a hydra happens to settle in gently flowing water which keeps its body moving constantly to and fro, the stimuli from the water gradually lose any elicitatory effect, and the hydra leaves its body and arms fully extended, passively following the movements of the water. But at the same time — and this is the important point — the threshold of all the other stimuli that cause contraction remains unchanged. This would certainly not be the case if the stimuli from the current did not completely lose their effect, but continued to provoke the animal to occasional contractions, however slight; then the motor side of the response would gradually weaken and the capacity to respond to all other stimuli would lessen. This is precisely what is prevented by habituation.

Habituation can also be called 'desensitation'. The term 'sensory adaptation' is misleading in a further sense inasmuch as it suggests that the processes in question take place in a sense organ, as, for example, when the retina of our eye, or the size of the pupil, adapts itself to the lighting conditions. Going from a brightly lit room into the darkness outside, one says that one 'must get used to the darkness'. The concept of habituation in our present sense, however, only rarely applies, as in the case of the eye, to processes attributable to changes in the sense organ itself, but to those that occur in the central nervous system. These processes are also generally longer in duration than those of sensory adaptation proper.

Margret Schleidt has studied habituation in the gobbling of turkeys and has proved that it is not a process that takes place in the sense organ. The gobbling can be elicited by many kinds of sound, and if a short note of constant pitch is emitted at intervals by a sound generator the turkey will first make the gobbling noise at every one of these stimuli, then less frequently and finally not at all. If one then produces notes of different pitches, one finds that the desensitation only affects a very narrow range of notes above and below the original note. The 'adaptation curve' drops away sharply from the peak on both sides, while the threshold of notes only slightly further removed in the scale from the original note is not affected.

Down to this point one could still have assumed that adaptation or a process of fatigue had taken place in the organ itself. That this is not the case, however, was shown by Margret Schleidt by an impressively simple experiment. She reproduced the same note as before, at the same pitch and with the same duration, but much more softly. To everybody's astonishment this elicited a full response again, as though it were a completely different note. The desensitation could not therefore have taken place in the sense organ, for in its adapted or weakened state it would have responded even more weakly to the softer note than to the preceding louder one.

From observations made in natural circumstances it also becomes clear how closely the waning of the original response is connected with a particular combination of external stimuli. The very complexity of such conditions shows that the higher functions of the central nervous system must be involved. Many anatids, for instance, react to carnivores stalking about on the banks of their stretch of water by 'mobbing' them, e.g. by emitting warning cries, and keeping the enemy in sight as long as possible. This response is elicited especially by the fox, or indeed any object with a russet coat. Dutch wild fowlers have exploited this in a particularly cunning way: they tie a fox pelt to the back of a trained dog, which entices the ducks into a long spiral channel leading to the trap. When we transferred our colony of waterfowl to the Ess-See, where there was no fence to keep out the foxes, we were afraid that because they were habituated to my Chow-Alsatian crosses, which have russet coats and somewhat resemble foxes, the birds might be in danger. They did in fact allow the dogs to come so close that, if they had been foxes, the ducks would not have survived. Our fears proved groundless, however, for the waning of the response only applied with our own dogs; even a neighbour's chow elicited the birds' mobbing, and foxes still more so.

It is often a matter for wonder what slight variations in the context of a stimulus situation can suffice to cause the collapse of a habituation to it. It was enough, for example, for one of our dogs to appear on the opposite bank of the lake to arouse the full mobbing response in our ducks and geese. I experienced the same situation with Shama thrushes (*Copsychus malabaricus*). A pair of these birds that had nested and bred in my room drove the fledglings of their first brood away at the moment when the next brood was close to fledging. A few days before this event I put a young male bird in a cage to protect it — principally from paternal attacks. However, the parent birds habituated themselves to the presence of their non-dislodgeable son by the process, as it were, of 'creeping in' of stimulation, ignoring both the cage and its occupant. But when, without thinking, I moved the cage to a different part of the room, the 'adaptation' was completely destroyed, and both parents attacked the young thrush through the bars so persistently that they completely forgot everything else, including the young of their other brood. And as these youngest birds could not yet eat by themselves they would have starved if I had not taken the offending cage and its inmate out of the room.

One puzzling aspect of the phenomenon of habituation is that in many cases the 'adaptation' process, so far from serving the slightest practical purpose, appears to be of negative survival value. We are familiar with a number of very specific responses which, although obviously serving the interests of survival, become so quickly desensitized that their full effect is only felt with their initial release, as Hinde has demonstrated from the warning- and escape-response which the owl elicits in the chaffinch. Even after not being elicited for a period of several months, the response did not attain anything like its original intensity nor did the strongest of possible conditioning stimuli, namely when an actual little owl was allowed to chase the chaffinch and even to pluck out a few of its feathers, have the anticipated effect of reviving the response. It is hardly conceivable that so specific and highly developed a mechanism, so clearly evolved phylogenetically to perform a particular function, should be there just to operate on one, or at most two, occasions in the lifetime of the individual. There must be some flaw in our reasoning or our experimental methods. In our experiments with greylag geese the response to the simulated warning call of their parents waned as rapidly as that of the chaffinch to the owl in Hinde's experiments, and never recovered. Possibly we destroy the response by being too impatient, seeking to elicit it too often and to repeat it too soon; or possibly we are encouraging an abnormally rapid adaptation by allowing our controlled experimental conditions to produce a uniformity of conditions which does not exist in natural life.

Wolfgang Schleidt has investigated a case in which desensitation actually conveys adaptive information. Turkeys threatened by a predator display an escape response elicited by a very simple pattern of stimuli. Anything that stands out as a dark silhouette against a bright background and moves at a certain rather slow angular velocity related to the object's length appears to the turkey as a predator, be it a fly crawling slowly across a white ceiling, a buzzard hovering overhead, a helicopter or a balloon. When Schleidt compared the relative effectiveness of different forms of stimulus — a goose in flight, for instance, or an eagle — it emerged that the form itself was immaterial but that habituation worked so quickly that the stimulus which remained most effective was that which had been withheld from the turkey for the longest period of time. In the field the strongest 'flying predator response' displayed by our turkeys was to a small advertising balloon that flies over our area once or twice a year; they responded far more weakly to the helicopters that fly over much more frequently, and weakest of all to the buzzards that circle overhead almost every day. The information given to the bird thus amounts to 'Beware of objects hovering above you in the sky, especially the rarest'. In natural conditions in North America this would mean the white-tailed or bald eagle (*Haliaetus leucocephalus*), the only bird of prey that can threaten full-grown wild turkeys.

As mentioned above, the process of habituation or desensitation differs from the most elementary processes of the modification of behaviour, i.e. motor facilitation and sensation, in one essential respect — it is accompanied by the process of association, which links the innate releasing mechanism to highly complex processes of Gestalt or form perception which will be discussed in a later chapter. This link has an inhibiting effect, the physiology of which is still something of a mystery. In their accustomed stimulus situation, which may be characterized by a highly complex combination of stimulus data, the innate key stimuli lose their irritating effect, but in all combinations with different ştimuli, however slight the differences may be, they retain their effectiveness.

5.6 *Habit*

'To accustom or adapt oneself to something' or 'to form a habit', in everyday parlance, means not only the process by which we get used to an annoying stimulus, thereby putting it out of our minds and rendering it ineffectual, but also when we become accustomed to a certain repeated stimulus situation or mode of behaviour in such a way that we come to like it, even to the point at which we cannot do without it. Just as with habituation, so here too a firm 'association' occurs, linking the key stimuli, to which a releasing mechanism responds, with the complex of environmental stimuli that accompanies them. As a result, the response that could originally be elicited by the simple configuration of the key stimuli alone now requires the whole configuration of all stimulus data, innate and acquired alike. In this case association thus has the exactly opposite effect from that in desensitation as described in the preceding section. In this latter it renders the original key stimuli ineffectual; in the process of 'accustomation' or habit formation, however, not only do the key stimuli remain effectual but they only develop this effectuality in conjunction with the situation to which the organism has become accustomed.

The survival value of this process lies in the great increase of selectivity in the releasing mechanism, and, unlike habituation, examples of this are especially to be found in the higher animals and man. An old caged bird that has eaten for years from the same bowl may starve if this bowl gets broken and he is expected to feed from a different one. The process can assume a pathological character with senile human beings whose behaviour becomes utterly deranged if there is the slightest change in their surroundings.

The survival value of this habit formation is at its most apparent in the ontogenetic development of certain animals. A newly hatched greylag gosling, for instance, 'greets' and then runs after any object which moves and responds to its 'distress signals' with rhythmic sounds of medium pitch. Having reacted once or twice in this way to a human being it can hardly be persuaded any more to follow a goose or a dummy; and if, after a great deal of patience, one finally succeeds in making it do so, the bird never shows the same intensity and fidelity that it showed towards the first object. This irreversible fixation, known as 'imprinting', will be the subject of a later section.

This imprinting of the gosling's response, whether it be linked to a man or a goose, related initially only to the species, not to an individual. A goose-imprinted gosling that is just able to run can easily be transferred from one goose family to another. But if it has been following its own parents for, say, two complete days, it can identify them as individuals — rather earlier by their voices than by their appearance, although, odd as it sounds, they recognize each other — like human beings — primarily by their facial features. They are completely unable to recognize individual conspecifics unless they can see their faces.

The gosling becomes selectively accustomed to its parents without the influence of any conditioning or deconditioning factors. Goslings sometimes lose their parents temporarily in the early stages and try to attach themselves to another pair of geese and their goslings, but the outsider is usually attacked and driven off. Such unpleasant experiences, however, do not prevent the gosling from repeating its mistake, or, if it finds its way back to its parents, makes it stay closer to them in future. Indeed, it is as though by following a strange pair of geese, even for a short while, the memory of what its own parents looked like seems to get blurred, and a gosling that strays away from its parents once is likely to do so again and again.

Or, to take another example: a two-month-old baby whose motor pattern of smiling is beginning to mature can, as René Spitz has shown in precise experiments, be made to smile by confronting it with very simple dummies. All that is necessary is the configuration of two eyes with a sort of nose between them, and for the dummy to appear to nod its head. The effect is strengthened by a well defined hairline, which makes the nodding more conspicuous; a mouth with its corners drawn upward in a grin forms an additional key stimulus. At first a balloon crudely daubed with these features proved to have the same effect as when a nurse nodded and smiled at the baby. But a few weeks later, during which the baby had smiled at real people much more often than at dummies, the dummy quite suddenly lost its effect. The baby had learned 'what a human looks like' and was now afraid of the painted balloon he used to smile at, although — and this must be emphasized — the balloon was not associated with anything unpleasant done to him which might have had a deconditioning effect.

Considerably later, at an age between six and eight months, the smile-releasing mechanism very suddenly becomes much more selective. From then on the baby smiles only at its mother and a few other people it knows well, shrinking away from everyone else.

With the faculty of individual recognition that of human bond formation begins to mature. It can have tragic consequences if, as still happens today with the continual changes of personnel in hospitals and kindergartens, a child is deprived of the opportunity to make the releasing mechanisms of its social behaviour progressively more selective and thus develop social ties with particular people.

This shrinking back from strangers on the baby's part must be due to an accustomation to individuals independent of any deconditioning experience through contact with strangers. In fact, the fewer strangers the baby sees, the more intensely it shrinks away from them.

5.7 *Avoidance responses acquired by 'trauma'*

I now come to the discussion of a learning process which the majority of psychologists equate with the acquiring of a conditioned reflex. However, I believe that it is a much simpler phenomenon and does not need to be explained in terms of the complex feedback mechanism of the conditioned response, to which I shall turn in the following chapter.

Even after a single occasion, a key stimulus that elicits an innate escape response of maximum intensity often becomes inseparably linked with the overall stimulus situation, both as it is at the time and as it was immediately beforehand. Of all forms of association this is the one that is found at the lowest level and is probably linked by a series of intermediate stages to simple processes of sensitation. In certain flatworms, for instance, a light stimulus which may of itself have only a subliminal effect on the escape response, may become effective by association with a strong 'unconditioned' escape-eliciting stimulus — a process regarded by many American students of behaviour as one of conditioning. In lower invertebrates with no centralized nervous system all such allegedly conditioned responses are acquired by processes of this kind, and the sum of their learning consists simply of this process together with that of habituation described above (p.71) in respect of the hydra.

Like habituation, the acquisition of escape responses in advanced creatures is linked with the whole function of complex Gestalt perception. A dog that had once been caught in the revolving door of a hotel entrance not only avoided all revolving doors from then on but also kept away from the scene of the trauma. Whenever he was made to go down the same street again, he would cross to the opposite sidewalk before reaching the hotel door and then rush past with his ears back and his tail between his legs.

'Psychic traumas' of this kind, as psychoanalysts call them, establish a virtually irreversible association between a complex stimulus situation and an escape response, as dog-trainers and horsemen know only too well, and a single such experience may 'ruin' an animal for ever.

5.8 *Imprinting*

The irreversible fixation of a response to a situation which the individual may encounter only a few times in his life can also come about through the process known as imprinting, or object fixation. From the physiological point of view the remarkable thing about this process is that the permanent association between the behaviour pattern and its object becomes established at a time when it is not yet functional — in many cases, indeed, not yet even traceable. The sensitive period for imprintability often belongs to a very early stage in the development of the individual; sometimes it is confined to a mere matter of hours, but it is always relatively well defined. Once it has been fixed, the determination of the object (see p.22) cannot be reversed. Thus animals that have been sexually imprinted on different species will remain incurably 'perverted'.

The majority of known imprinting processes concern social behaviour, such as the following response of young nidifugous birds, rival fighting, common to many bird species, and especially sexual behaviour. It is misleading to say that this bird or that mammal is imprinted — human imprinted, for instance — for what is determined in this way is only the object of a narrowly specific behaviour pattern. A bird fixated on a different species as concerns sex certainly does not have to be so in other respects, such as fighting with rival birds, or other forms of social behaviour. It is fortunate for our investigations that in the greylag goose the following response and other social behaviour patterns of the young animals are easily human imprinted without sexual imprinting occurring at the same time.

There are also cases in which the behaviour of parasites is imprinted on the species of their host. W. H. Thorpe showed, for example, that ichneumon wasps lay their eggs in the same species of caterpillar as that in which they were themselves hatched. By 'transplantation' of the larvae one can imprint wasps of a species which normally parasites bee moths, on flour moths. Bruns has demonstrated in ants that each individual fixates its social response on the particular species of ant which helped it hatch from the pupa. This phenomenon is at the root of the so-called slave-holding of some species of ants. Monika Holzapfel has shown that the behaviour of owls when they catch their prey is imprinted on a particular species of victim, and that if the sensitive period for imprintability is wasted, an individual owl may become incapable of ever catching prey at all.

Imprinting is also connected by some intermediate stages to a variety of other processes of associative learning. When the young of certain passerine birds learn their species-specific song by imitating adults (as M. Konishi has shown), the learning process is restricted to a sensitive period, just like typical imprinting processes, and is irreversible. Intermediate processes of this kind have caused misunderstanding. Hinde, Bateson and others have investigated processes which differ considerably from those of typical imprinting, such as the ways in which domestic chicks attach themselves to their mother or to some substitute object. Such processes are closer to normal learning processes than to imprinting. On the basis of such investigations doubts have been expressed on the observations reported by Whitman, Heinroth and me, but the work of Immelmann, Schein, Konishi, Schutz and others has fully confirmed what was discovered more than twenty years ago.

As is the case in habit and habituation, imprinting is also 'associated' with complex perceptual operations, and, as with those two processes, it is learned within the framework of an innate release mechanism. Consequently this mechanism is rendered more selective by the imprinting process.

One of the most interesting and puzzling functions of imprinting is that the imprinted response does not become fixated on an individual but on a species. The sexual responses of a mallard drake reared in the company of a shelduck (*Tadorna tadorna*) are not imprinted on this one individual but on the species. Given a choice of shelducks, the mallard hardly ever chooses its 'imprinting partner' — this is prevented by incest-inhibiting mechanisms — but another bird of the same species. A jackdaw I had reared and which had become sexually human imprinted, directed its courtship behaviour towards a petite dark-haired girl. What caused the bird to take both of us as belonging to the same species is a puzzle I have never been able to solve. Thus the perception of the releasing stimulus combination performs a function analogous to abstraction.

Another unsolved question is whether certain conditioning stimuli of the rewarding and hence trainable type play a part in the process — in other words, whether imprinting is to be conceived of as a conditioned response in the Pavlovian sense. Against this view is the fact, mentioned above, that the imprinted object is often already firmly determined before the animal has performed the particular behaviour pattern or shown the slightest sign of doing so. A jackdaw, for example, is sexually imprinted shortly before it leaves the nest, and up to this moment it cannot possibly be said to have been sexually aroused even to the slightest degree. Two years have to elapse before its copulatory urge begins to mature, which, as a consummatory act, should be expected to be the most important of its conditioning forces. However, this does not entirely rule out the possibility that other conditioning stimuli may also be involved, though they have not yet been identified as such. On the other hand, nothing compels us to make this assumption, and in all probability imprinting is an associative learning process analogous to those described in the preceding two sections. In its irreversibility and its restriction to limited phases of ontogeny, imprinting, more clearly than all other learning-processes, bears the stamp of what Spemann called *induction*.

5.9 Summary

Chapter 4 dealt with physiological mechanisms which acquire and exploit instantaneous information but do not store it. They can all function an unlimited number of times without any change being caused in their machinery; indeed, being the foundation of all potential experience, they have to be resistant to any modification through experience.

This chapter has dealt with fundamentally different processes in which behaviour mechanisms are modified in the course of the individual organism's life — modified, moreover, in a way that improves their survival function.

For the survival value of a structure and its function to be increased by modification is no more likely than if this were to be achieved by genetic mutation. Whenever particular external circumstances regularly produce modifications which result in an adaptation to these very circumstances, there is an overwhelming probability that we are dealing with what Ernst Mayr called 'open programmes' specified in the genome.

A genetic programme of this kind contains several individual programmes for the construction of various mechanisms and therefore presupposes not less information than one single closed programme, but far more. On the other hand, the open programme has the capacity to absorb further information from outside by allowing this information to determine which of the available possibilities shall be realized. When this has been done a new adaptation is made permanent, and the information on which it is based thus becomes stored. In this way the central nervous system repeats on a higher plane a function present in the genome but missing from the processes by which instantaneous information is acquired.

All learning is an adaptive, teleonomic modification of physiological mechanisms whose operation constitutes behaviour.

Experimental embryology offers a good illustration of open programmes and adaptive modification. Which of the 'prospective potencies' of an embryonic tissue is utilized depends on environmental influences, and processes of adaptive modification, including those of learning, are related to induction, as defined by Spemann.

The most elementary forms of adaptive modification of behaviour are facilitation on the motor side and sensitation on the receptor side. The latter processes only have survival value when there is a high probability of the elicitatory situation arising a number of times in succession.

All the other processes of modification discussed in this chapter depend on an 'association', on the forming of a linkage between two nervous processes hitherto not causally connected. As a result, often very complex stimulus situations can come to influence very simple innate behaviour patterns.

In the process of habituation this influence has an inhibiting effect: as a result of association the original releasing key stimuli lose their effectiveness, but retain or regain it with even the slightest modification of the situation as a whole. As every good cook knows, variety reawakens the waning response to certain key stimuli.

In the reciprocal process of accustomation or habit formation key stimuli are linked with a complex stimulus pattern in such a way that from then on they only retain their effectiveness in combinations; as a consequence, the selectivity of an innate releasing mechanism greatly increases.

In the case of powerful escape responses the releasing key stimuli often become associated with the accompanying stimulus situation after a single deeply impressive 'traumatic' event; whereupon there develops a strong escape response to a complex, hitherto ineffective stimulus situation. This association is frequently irreversible.

Many behaviour patterns, especially social patterns, become irreversibly fixed on an object at certain early, sensitive phases of their development. By virtue of its irreversibility and its restriction to a sensitive period, this 'imprinting' is closer than any other learning process to what Spemann called 'induction' and 'determination'.

The learning processes described in 4-7 of this chapter are usually subsumed under the concept of association. They establish a new link between independent nervous processes. The general ideas about learning held by earlier psychologists such as Wilhelm Wundt and C. L. Hull correspond very closely to these processes. On the other hand a critical comparison of the various American theories of learning, such as that undertaken by C. Foppa (*Lernen, Gedächtnis, Verhalten*, 1966) makes abundantly clear how often these theories suffer from a tendency to seek a single explanation for processes basically different in their causation (see p.39). Attempts are made time and again to treat all learning processes in terms of a single comprehensive theory. But what is here called 'learning' is in fact an imaginary halfway house that lies between the processes discussed in the previous chapter and those that are based on a completely different and more complex organization of nervous processes. These are the subject of the following chapter.

Chapter six

Feedback of experience: conditioning by reinforcement

6.1 *New feedback*

All animals whose central nervous system has reached a certain level of differentiation — cephalopods, crustaceans, arachnids, insects and vertebrates, including man — possess a faculty for acquiring knowledge which surpasses all the cognitive mechanisms hitherto discussed — namely the faculty of learning in the strict sense of the term. That this faculty is present in so many different creatures misled psychologists unfamiliar with biology or convergent adaptation into thinking that it was an *ur*-phenomenon — the basis of all knowledge, indeed the sole factor in behaviour of any kind. In fact, however, the above five phyla developed by adaptation the nervous processes underlying their powers of learning as independently as they developed their eyes and their extremities.

Learning through trial and success, as a characteristic *fulguratio*, or 'creative flash' as described above (p.29f.), arose as a new combination of separate nervous mechanisms already functioning independently of each other. The unitary behavioural mechanism termed species-specific drive activity (Heinroth's *'arteigene Triebhandlung'*) consists, as we already know, of appetitive behaviour which achieves the stimulus situation to which an innate releasing mechanism specifically responds, and of a phylogenetically programmed behaviour pattern which assuages the motivation and achieves the 'satisfying' terminal situation. This chain, consisting of three separate processes, represents the base on which all learning through success and failure, i.e. conditioning in the strict sense of the word, has evolved. This linear sequence of processes acquires unexpected and truly epoch-making system characteristics by the 'invention' of a feedback loop through which the final success or failure of the chain of processes is able to have a modifying effect on its initiating links.

As a result of this feedback the elements of the appetitive behaviour, formerly more or less random in their appearance, are reinforced if the survival-purposive aim of the behaviour is achieved, but weakened if it is not. Put in another way: success acts as what one generally styles 'reward', failure as 'punishment'. Everything that promotes the former is designated as 'reinforcement'. The actual concept comes from Pavlov. When I asked a Russian-speaking colleague of mine to find out where Pavlov first used a term to define this concept, she discovered that his early works, where it first appears, had been written in German, and that his term was *Verstärkung*, 'strengthening'. This term is not very satisfactory. The word which, in my opinion, best describes what is happening is 'encouraging'. Success encourages the behaviour pattern that promotes it.

This new feedback leads to a cognitive process which gives the individual twice as much knowledge in a single run-off of the pattern as the genome, even in the most favourable circumstances, could do in a whole generation, because it derives information not only from success, like the genome, but also from failure. Moreover, unlike the genome which chooses blindly from among both relevant and irrelevant factors, the learning process proceeds on the basis of well-tried, innate working hypotheses which are firmly established in the behavioural systems of all advanced animals in the form of mechanisms for acquiring instantaneous information (see Chapter 4). Thus behaviour that can be modified by success and failure tends from the outset towards a higher probability of success. That the Latin words for 'from the outset' are *'a priori'* is certainly not a coincidence. The great efficiency of this new cognitive apparatus explains why, among advanced animals with a high degree of mobility, only those that possess this apparatus can compete for survival.

6.2 *The minimum complication required*

It is clear from what has been said why the process of 'learning by success' cannot evolve in unicellular or lower multicellular creatures which have no centralized nervous systems; for a system that is capable of exploiting the success or failure of a particular behaviour pattern as a source of knowledge, and of using this knowledge as feedback to achieve an adaptive modification in the machinery of that pattern, obviously assumes the existence of various complex and highly organized subsystems. One of these subsystems is the species-specific drive activity discussed on p.55.

The easiest to imagine would be the kind of mechanism that reinforces a behaviour pattern designed to satisfy simple tissue needs. All that would be required in this case is a single 'feeler' which registers the presence or absence of a particular indispensable substance, and transmits this information to the apparatus of the earlier behaviour. Individual cases of this very elementary example of a potential conditioned response have actually been found; V. G. Dethier demonstrated that certain flies, for example, obtain their food in this way. In general, however, and in the majority of behaviour patterns susceptible to adaptive modification by learning in the true sense, the following three conditions must be satisfied:

(1) The behaviour pattern that initiates the action must have a 'wide open' programme, one that offers scope for manifold adaptive modifications. As we know, a programme of this kind presupposes a particularly large volume of genetic information.

(2) The circumstances and the manner in which the initial links of the chain of activities have been performed in each individual action must be recorded in some sort of memory, and this record must be brought into relation with the success attained in this particular case.

(3) The report concerning success or failure which is fed back into the machinery of precedent behaviour must be reasonably reliable. The consummatory act, as Wallace Craig called it, or in the case of what Monika Meyer-Holzapfel termed appetitive behaviour striving for quiescence, the relief of tension, must be characterized by internal or external receptor processes to a degree which makes erroneous reports sufficiently improbable. In other words this receptor mechanism must achieve a function similar to that of the innate releasing mechanism (see p.53 *f*.). Any simpler model of the physiological mechanisms required for conditioning

by reinforcement is inconceivable.

A neural organization performing that kind of function can therefore never be a simple 'reflex' in the sense implied by Pavlov's term. To be sure, there are straightforward escape responses like those discussed in Chapter 5.6 (p.74f.), which are produced by simple association of response and elicitatory stimulus situation, and which bear a superficial resemblance to the learning processes here under discussion. But at the same time we know of no single case in which a behaviour pattern could have been adaptively modified by means of positive conditioning stimuli, for example reinforcing stimuli, if appetitive behaviour were not also involved. Tolman pointed this out long ago.

Even in Pavlov's classic case of conditioning salivation in a dog, it is not merely excretion of the saliva that is being reinforced by learning. The dog's secretion of saliva is only a small part of a far more complex behavioural sequence. In the standard laboratory experiment, however, most of the elements in this sequence are eliminated, because the dog is held so tightly in a leather harness that it can scarcely move. When my late friend Howard Liddell was working as a visitor in one of Pavlov's laboratories, he aroused indignation by performing an unorthodox experiment. First he trained a dog to respond by salivation to the *acceleration* in the beat of a metronome. When the response to this conditioning stimulus was reliably established Liddell set the dog free. It immediately ran towards the table where the metronome was still ticking at a constant rate and fawned on it, wagging its tail and whimpering — in other words behaving just like a dog begging its master for food. It secreted a great deal of saliva the whole time even though the metronome had ceased accelerating. Begging and mutual feeding are common among social canidae: L. Crisler showed that wolves only one year old feed others younger than themselves and African hunting dog (*Lycaon*) feeds all the animals in the pack with the prey that it catches. In both cases the motor patterns of begging are the same as those of the domestic dog, and it is these patterns, rather than just the secretion of saliva, that epitomize the response elicited in Pavlov's experiment.

It is far from my intention to disparage Pavlov's experiments, and it is entirely justifiable to isolate a single response artificially, especially if, as in the case of Pavlov's dogs, the experiments provide particularly good opportunities for quantitative analysis. But one must never forget that one has isolated part of a system, and must beware of falling into the common error of believing that the part is the whole, sufficient in itself to reveal all the characteristics of the whole.

If, from the viewpoint of a systematic analysis, one surveys the facts today known about conditioning by reinforcement, one feels definitely confirmed in the opinions expressed in the above. Positive learning by rewards is the most important criterion of 'genuine' conditioning. Among 'conditionable' responses listed by C. Foppa, there are some in which avoidance responses brought about by mere association simulate true conditioning.

6.3 *In search of the engram*

Before turning to the question of which parts of a system amenable to modification by learning are actually affected by adaptive modification, and where the new information comes from, we should make a few general observations on the physiology of learning and memory.

Attempts to discover the engram, the residual effect of an experience in memory, have so far been almost discouragingly unsuccessful. In 1950 K. S. Lashley gave his fascinating talk, 'In Search of the Engram', subtitled 'Thirty Years of Frustration'. But among the extremely important conclusions to which Lashley's patient researches led, was the discovery that the engram is not located in one particular part of the brain but consists of an organization linking a mass of different parts of the brain. Even today we are not in a position to say what the physiological processes behind this organization are. This explains why, when it was discovered that genetic information was encoded in the chain molecules, many reputable scientists immediately advanced the hypothesis that the knowledge one gains from personal experience and stores in one's memory is governed by the same process.

There are, however, serious objections to this hypothesis. For it to be correct, there would have to be two independent mechanisms: one would record all the nervous impulses received and arrange them, much as on a tape recorder, in a sequence of nucleotides encoded in the chain molecule; the other would have to be able to read this code and convert it back into a pattern of nervous impulses coordinated in time and space. Apart from its basic improbability, this hypothesis cannot explain why in all known animals learning capacity stands in direct ratio to the number of ganglion cells — in fact, to the size and degree of sophistication of the central nervous system. Biochemists have recently shown that there is not time enough for the chemical encoding of individually acquired information in chain molecules. Moreover, further critical examination showed that many of the results which seemed to show that individually acquired information had been chemically transferred, were incapable of being repeated. I therefore still adhere to the belief that all learning functions, at least in so far as they involve complex adaptive modifications of behaviour, occur at the synapses, i.e. the conjunctions, of the individual neurons, and that these modifications are closely related to the embryogenetic processes of induction. At the same time, of course, I do not deny that changes in the molecular code may also play a part in this process.

6.4 *Innate teaching mechanisms*

The open programme of behavioural mechanisms inherited by every individual through the phylogenetic development of his forbears is always so constructed that its variable components relate to environmental situations whose nature and occurrence in time and space are unpredictable; however, they are sufficiently constant factors in the individual's life to be worth storing information about them. A newly hatched greylag goose cannot possibly know what the parents look like which it will have to follow in the months to come, nor can a young bee possess inborn knowledge about the surroundings of its hive. The ability to recognize individual conspecifics and a susceptibility to path conditioning are good examples of the faculties needed to acquire relevant knowledge which cannot be provided either by the genome or by the mechanisms for acquiring instantaneous information.

On the other hand, we have seen that an open programme of this kind presupposes a large body of phylogenetically acquired, genome-bound information. This information is converted into relevant behaviour, but not through morphogenetic development — although in the first instance it is morphogenesis which, on the basis of such information, causes particular neural mechanisms to develop, such as the process described on p.77f. Similarly, the whole structure of the reinforcement apparatus described in this chapter on p. 85f. depends on information from the genome. If one confines oneself to natural explanations, one cannot conceive that information from the genome can be converted into survival-purposive behaviour other than by the development of actual physical structures of sense organs and nervous system.

It is these structures that direct learning into profitable channels and produce the teaching mechanisms which ensure that the gaps in the various programmes are filled in relevant ways. As has been said many times before, these structures themselves need to be as little susceptible to modification as possible, so as not to lose the innate information they contain. If one part of a behavioural system can be considerably modified by learning, one is bound to assume that other parts are sufficiently resistant to modification to ensure that the learning programme of the variable parts is carried out.

Unless one believes in supernatural factors, such as a pre-established harmony between the organism and its environment, one has to postulate the existence of innate teaching mechanisms in order to explain why the majority of learning processes serve to enhance the organism's fitness for survival. These mechanisms also meet the Kantian definition of the *a priori*: they were there before all learning, and must be there in order for learning to be possible.

It is a fascinating undertaking to explore a complex behavioural system which requires adaptive modification by learning in order to function, in search of those open programmes which, though genetically determined in themselves, are the organs acquiring experience. These 'innate teaching mechanisms' are of the most varied kind and can be concealed in sense organs as well as in the nervous system. They can, for instance, be concentrated in receptor mechanisms. In what Wallace Craig called 'aversions' and Monika Meyer-Holzapfel, more appropriately, the 'desire for quiescence', phylogenetically programmed receptor processes inform the organism when something is 'not right' in its environment, such as when it is too dry or too wet, too hot or too cold, too bright or too dark, when the biotope does not afford enough cover or contains too many obstructions to visibility, etc. The kinetic agitation that grips the animal as long as the aversion-producing situation lasts can assume the most varied forms and levels, from the simplest kinesis to complex aiming patterns involving learning and insight. Wherever it occurs, the adaptive modification always affects the patterns of pathfinding and genuine conditioned responses are produced — so-called path habits — which remove it from any harmful stimulus as quickly as possible.

Another way in which conditioned responses arise, equally common and with an equally simple innate programme, serves the important function of preserving by exterior behaviour the constancy of conditions within the organism, i.e. maintaining the various homeostases by means of relevant responses. We know from our own experience that we receive reliable information if something in one of the various cycles in our body is not working as it should do. Sometimes the information is specific, as with hunger or thirst, or when cell tissue is deprived of certain substances. The earliest behaviourists, such as Thorndike, took the view that the satisfaction of tissue needs was the most important reinforcement of the self-conditioning process. But they did not ask how the organism as a whole, and in particular its central nervous system, could 'know' what it was deprived of, or how it could remedy the situation.

Another, slightly different, example of a mechanism that transmits information on disorders is our sense of pain. This particular function is to locate the discomfort: we learn immediately where it is, and are not allowed to forget it. Of particular interest, however, are the least localized messages which our body conveys to us about malfunctions in its cycles, where we can only say that we 'feel bad'. In the case of a slight infection, for instance, we are quite unable to indicate the source of the discomfort, however miserable we feel. If, however, we feel like vomiting, we have specific local information and may even be in a position to say why we feel unwell. This is clearly the source of the survival value of the information. If our discomfort is the result of eating bad food, we generally find ourselves thinking — in 'free association', as psychoanalysts put it — of something slightly suspicious we ate the previous day, which then makes us sure that we have identified the cause. Avoidance responses caused by such an experience can persist for a long while, often even for a whole lifetime.

The innate teaching mechanism that produces these conditioned responses by causing feelings of discomfort to act as 'punishment', and feelings of well-being as 'reward' or reinforcement, can be programmed in a very general way. All it needs is to have a 'feeler' in certain cycles in the organism, and to 'punish' changes that lead away from the desired norm, or *Sollwert*, and 'reward' those that move towards it. This is the principle behind the mechanism that makes many omnivorous animals decide what to eat. Many years ago Curt Richter discovered that when rats were given the various nutrients they required in a number of individual dishes, even to the extent of breaking proteins down into their individual amino acids, they took from each dish just what they needed for a balanced diet. Since a rat cannot possibly possess phylogenetically acquired information about which amino acids can be synthesized to form a wholesome protein or how much of each such acid is needed, it must get its knowledge from somewhere else.

Experiments performed by J. Garcia and F. R. Ervin have produced valuable results concerning the particular learning programme providing this information. They found that rats could be conditioned to accept or reject certain nutrients only by influencing the viscera. In order to induce nausea and vomiting they gave the rats injections of apomorphine or exposed them to X-rays for the period necessary to produce the same results. All attempts by means of pain stimuli and other extreme forms of 'punishment' to get the rats to refuse certain foods met with no success. Similarly it proved impossible to use intestinal stimuli to decondition any behaviour pattern other than that of ingesting certain foods.

Both in the adaptive modification of the urge towards a state of quiescence and the above examples of self-conditioning to certain foodstuffs, it is conditioned avoidance responses that play the most important role. Hence there is a certain justification for talking generally of 'aversions', as Craig has done. But if, for example, an animal crawls from a cold to a warm zone, we can never determine objectively whether it is escaping from the cold or looking for the warmth. This is why I prefer Monika Meyer-Holzapfel's term 'seeking a state of quiescence'. In both cases, it is clear that the organism has been irritated, and that it is the waning of this irritation that acts as the conditioning reinforcement. This is the crucial phenomenon of 'reinforcement by relief of tension', as identified by Hull.

There are also modifiable behaviour systems in which innate information is present not only in the receptor mechanism that analyses the stimulus situation but also in the fixed motor pattern itself. A good example of this is the way the jackdaw and other corvids build their nests. Standing in the centre of the chosen area, with the nest material in its beak, the bird makes a strange trembling movement as it pushes the material sideways and slightly downwards in a broad arc, pressing it against the base of the nest or the parts that are already built. If the bird strikes an obstacle, the trembling movement intensifies, and the pushing movement turns into a series of vigorous lateral thrusts, reminiscent of the way a man uses a pipe-cleaner when it meets an obstruction in the stem of the pipe. If the bird is carrying a twig or something similar, it laboriously pokes it further and further into the nest until it becomes lodged so firmly that it cannot be moved forwards or backwards. When this point is reached, the trembling movements come to a kind of orgasmic climax and suddenly stop. The bird has now lost all interest in the object and, for the moment, in building a nest. This whole process is a typical example of what Craig called a 'consummatory act'.

In contrast to many other passerine birds, jackdaws and other corvids appear to have no information in their release mechanism about the most suitable building material. When their urge is first aroused, they fetch the most unlikely objects and try to poke them into a suitable place — knowledge of what is a suitable *place*, at least, is innate. I have seen jackdaws and ravens trying to build nests with pieces of glass, the bases of old light bulbs and even lumps of ice. Such objects, of course, will not stay put, and the consummatory act is never elicited. But after a very short time the bird *learns* to use only such objects as produce, via the trembling movements, those reafferences which are programmed into the innate learning mechanism as reinforcing factors. This information is sufficient to teach the bird to choose material that can be used to build a firm nest. Occasionally the teaching mechanism makes a mistake usually caused by a dearth of information: thus wire or metal strips provide particularly strong reinforcing reafferences, and a bird may train itself to use this material, although, because of its heat conductivity it is biologically unsuitable. Nests containing metal are, indeed, by no means rare in factory areas. The process just described is a typical example of the effect produced by a so-called 'supernormal' object. Getting conditioned to such an object is equivalent to getting addicted to a vice. In the rat the learning processes by which the various action patterns of nest building are integrated are more complex. As Eibl-Eibesfeldt has shown, every single one of the motor patterns involved is entirely innate. So also is their sequence, for the rat 'knows' innately that building a nest has to start with collecting material far away from the nest and carrying it home. Rats reared by Eibl-Eibesfeldt in cages containing nothing that could be picked up used their own tails as substitutes, taking hold of them between their teeth some distance from where they were accustomed to sleep, carrying them home and carefully depositing them at the appropriate spot. Keen to experiment with entirely inexperienced animals, Eibl-Eibesfeldt repeated the exercise with young rats whose tails had been amputated long before the urge to build a nest had developed. When, having become older, they were offered strips of soft paper for the first time, they immediately began to build. Those who had chosen a particular place to sleep in the undivided cage at once put the strips of paper on that spot, those who before the experiment had been accustomed to sleep now in one place, now in another, took a few minutes to decide where to build their nest. When a corner of the cage was partitioned off by a piece of metal only a few square centimetres in area, it caused all the animals to build their nests within this protected area.

The behaviour of the inexperienced rats differed significantly from that of the experienced ones. For one thing, the intensity of performance was much higher: the animals pounced on the building materials with a greed readily explained by the fact that the action had been repressed up to that time. But the most important difference lay in the fact that the ordered sequence in which the experienced rats carried out the motor patterns of building their nests was missing.

To start with, an experienced rat confines itself to collecting material; not until it has collected a considerable amount will it proceed to form it in a circle around itself, building a kind of circular rampart round the centre of the nest. Only when this wall is high enough does the rat start the so-called 'papering movement', patting the inside of the wall with its front paws to make it smooth.

The rats in Eibl-Eibesfeldt's experiment carried out each of these behaviour patterns in a perfectly coordinated manner and no differences from the normal could be detected in any of them, even in the analysis of slow-motion films. However, the ordered sequence between these patterns mentioned above was completely lacking; the rats would busily arrive with a strip of paper, put it down and then perform the movements of circular building and 'papering' in empty space.

The system in rats is more complex than in jackdaws, but every one of the various innate teaching mechanisms present functions according to the same principle. In each case the conditioning and reinforcement of a specific sequence of actions is achieved by two processes: firstly by the fact that the form of the fixed motor pattern can only give the reinforcing feedback information in specific environmental conditions provided for by the programme, and secondly by means of feedback of success or failure provided by exteroceptor, and probably also proprioceptor mechanisms.

Moreover, there is likely to be an immediate deconditioning effect when a motor pattern fails to produce feedback of any kind. When one observes these learning processes at first hand, one gets the impression that the rat gets more satisfaction from its building activity when the building material is already on the spot, and that the 'papering' movement gives it complete satisfaction only after the wall round the nest has been erected.

The reader interested in learning theory may care to note that the learning processes in which a large proportion of the innate information is located not in the receptor area but in the fixed motor pattern itself, can be subsumed under the concept of operant conditioning. In this case, however, 'operant' refers not to a simple, multipurpose movement like scratching or scraping with the front paws, which could have a reinforcing effect by pure chance (like pressing the lever in a Skinner box or an old-fashioned 'puzzle box'), but to a highly specialized behaviour pattern capable of performing only one specific function — namely that for which it was phylogenetically evolved.

Elementary multipurpose behaviour patterns are governed by the laws laid down by Skinner for the learning process that he designated 'type-R conditioning'. The first of these laws states that the strength of the operant increases if it is followed by the appearance of the reinforcing stimulus situation, The second states that the strength of the operant decreases if, having already been reinforced by conditioning, it is not followed by the reinforcing stimulus. These laws apply if the operant is a so-called 'tool pattern', like locomotion or some other simple fixed motor pattern which, as multipurpose tools, can serve various goals or ends; they only partially apply, however, if the operant is a behaviour pattern where the animal is motivated by the drive for consummation. In this case, although the animal's earlier behaviour will in the long run presumably be reinforced by learning, for the moment it is totally obliterated; the drive has been satisfied, and the conditioning effect will only become apparent when the drive is aroused again. Furthermore, the absence of reinforcement does not make the animal give up the operant. Indeed, since the operant is a motor pattern in its own right, the failure to achieve consummation only leads the animal to try the harder to find an application for the pattern in other situations and with other objects.

It is important to make these distinctions, because the process of trial and error illustrated from jackdaws and rats, which is also found in basically the same form in many other animals, might be confused with the so-called 'exploratory' or 'curiosity' behaviour to be discussed later.

In the process described above the instinctive behaviour pattern is subject to its own inborn motivational pressure, and the animal tries to apply it in the same form with the most different objects. With exploratory behaviour, on the other hand, as Monika Meyer-Holzapfel has convincingly shown, the organism is quite differently motivated, in that, instead of finding one single behaviour pattern repeated with different objects, we now find a variety of patterns — often the animal's whole repertoire of action applied in succession to a single object. Both these learning processes differ from operant conditioning in its standard form, since here the operant is a generalized tool response which can be elicited by a wide variety of motivational pressures.

In his investigations into how young birds learn to sing Konishi came up with some particularly interesting and unexpected results concerning the location of innate information. It has been known for some time that in many species of passerine birds the young bird has to hear an adult of the same species sing if it is itself to develop a fully normal song. Then J. Nicolai revealed the surprising fact that certain birds, such as the bullfinch, only learn from specific individuals with which they have a close and equally specific social relationship. Young birds of many species, which have to learn by imitation, model their song on that of their own species, even though there may be a lot of other birds singing at the same time, and even if the voices of those of their own species are not the loudest or most distinctive. In addition, Heinroth observed that young birds which need to hear and imitate their fellows' singing but were reared in isolation, finally succeeded, after many attempts, in producing a song that was reasonably close to that of its species — a process that caused Heinroth to talk of 'self-imitation'.

All these phenomena were explained by what Konishi discovered. Birds whose auditory organs he had removed when they were very young later produced a vocalization that consisted more of noises than of notes, and was completely formless. This was true of species of which individuals left intact were reared in isolation in sound-proof chambers and developed a song clearly recognizable as that of their species. One is driven to the remarkable conclusion that these birds must possess what Konishi called an 'auditory template' of the song proper to their species: softly and in a playful way they try out highly varied combinations of sounds and retain those that correspond most closely to the inherited acoustical blueprint. The soft, cooing sounds they make at the beginning thus represent a kind of playful exploration of the situation, much as they do in a human baby.

The examples of phylogenetically programmed teaching mechanisms discussed above point to three significant facts:

(1) It can always be experimentally shown which subsystem of a complex, modifiable behaviour pattern contains the innate information, which ensures that an animal learns behaviour patterns required for its survival.

(2) It is not possible to understand any learning process without knowing the whole system which is modified by this process.

(3) One cannot make any universally valid statement about what constitutes reinforcement. Thorndike's theory that it is basically a matter of satisfying tissue needs, or Hull's notion that the essential factor is relief of tension, both apply only to particular cases. The physiological nature of the reinforcement process has to be studied independently in each individual case of learning.

6.5 *Modifiability of subsystems*

In the preceding sections, and more fully in my book *Evolution and Modification of Behaviour,* I have tried to show that it is impossible to assume that all individual component elements of behaviour are susceptible to adaptive modification, unless one is prepared to take refuge in a vitalistic belief in a pre-established harmony between the organism and its environment. Any kind of modifiability regularly leading to an increase of survival value inevitably presupposes a phylogenetically evolved open programme and a phylogenetically programmed teaching mechanism of the kind discussed in the preceding section. If there were complete fluidity in all behaviour patterns, one would have to assume an infinity of information and an infinity of teaching mechanisms — which would obviously be nonsense.

6.6 *Conditioned response, causality and transformation of energy*

As I mentioned earlier (p.12), cognitive mechanisms may often have arisen on different levels of integration through adaptation to one and the same fact of objective reality. Sometimes they may be distributed among animals of different levels of development, sometimes they may be found operating independently of each other in one and the same species. The same is true of the mechanisms to which I shall now turn my attention.

The most important function resulting from the ability to produce conditioned responses, as Tolman has emphasized in his book *Purposive Behaviour in Animals and Man*, is that it enables the organism to interpret a biologically irrelevant combination of stimuli as heralding the approach of a significant situation, and to prepare for the onset of this situation. I once watched semi-wild goats in the Armenian mountains make for the shelter of a cave at the sound of distant thunder, in order to protect themselves from the approaching storm. They reacted in exactly the same way to nearby blasting operations. I can remember very clearly now that when I noticed this I suddenly realized that in natural circumstances conditioned responses serve the interests of survival only if the conditioned stimulus stands in a causal relationship to the unconditioned stimulus.

It is only a causal linkage that, in nature, ever makes the *post hoc* follow with sufficient reliability to endow conditioning with a survival function. If, by pure chance, two unconnected events followed each other two or three times — which in some cases is sufficient to establish a conditioned response — the animal would be taken in by its propensity to become conditioned. A causal chain exists equally if the experimenter rings the 'dinner bell' before feeding the Pavlovian dog, though the chain of causation which acts within the experimenter's own central nervous system still defies our analysis. The 'hypothesis' underlying the mechanism of conditioning can thus be formulated: any *post hoc* that has occurred a number of times in the same exact sequence can be safely assumed to be a *propter hoc* — in other words, the effect of causal linkage.

This teleonomy of conditioned responses throws light, I believe, on a particular error in the empiricism of Hume to which Karl Popper draws attention in his *Objective Knowledge*. As Hume shows, it is impossible from a purely logical point of view to conclude from a number of precedents, however great, that a particular sequence of events must necessarily repeat itself, or even that the probability of this being so must increase as the number of repetitions grows. Hume then asks the psychological question: How does every reasonable man come to confidently expect that the sun will rise again tomorrow, that when he drops a stone it will fall to the ground, and that all the other events in the world will go on happening as they have before? Hume's answer was that it was a result of custom or habit. In other words, the fact that something happens repeatedly establishes an association of ideas, without which we could not manage to survive.

As Popper shows, the contradiction between logic and common sense has not only led many thinkers to despair of the possibility of ever attaining to objective knowledge but also compelled Hume himself to adopt a non-rational epistemology. In Popper's words:

> His result that repetition has no power whatever as an argument, although it dominates our cognitive life or our 'understanding', led him to the conclusion that argument or reason plays only a minor role in our understanding. Our 'knowledge' is unmasked as being not only of the nature of belief, but of rationally indefensible belief — of an irrational faith.

Let me quote just two sentences from the clear reasoning by which Popper shows the way out of this tangle. Even taken out of context they bear witness to the fundamental agreement between the conclusions of logic and ethology. Popper writes:

> I regard the distinction, implicit in Hume's treatment, between a logical and a psychological problem as of the utmost importance. But I do not think that Hume's view of what I am inclined to call 'logic' is satisfactory. He describes, clearly enough, processes of valid inference, but he looks upon these as 'rational' mental processes.

It is one of Popper's basic principles to convert all subjective terminology into objective terminology whenever problems of logic are involved. 'What is true in logic is true of psychology', he states quite simply. This principle of transference between subjective and objective corresponds precisely to my conviction, stated in the Prolegomena, that cognitive processes are identical with physiological processes.

Logical thought, like the formation of conditioned responses and numerous other 'psychological' processes, is a function of man's cognitive apparatus, the relationship of which to external reality has already been discussed (p.8f.). The bitter consequences of Hume's empiricism that all human knowledge is in reality nothing but an act of faith, without foundation, would only be correct if the old adage were also true, *'Nihil est in intellectu quod non antefeurat in sensu'*: 'There is nothing in our minds that has not already been in our senses.'

We know, however, how false this saying is, for all adaptation is a cognitive process, and our cognitive apparatus, which is what makes it possible for us to experience things, presupposes a vast amount of phylogenetically acquired information stored in the genome. Hume did not know this — and modern behaviourists do not want to know it.

After experiencing something a number of times, we all feel driven to look for some kind of connection between the experiences, however vague. I can still remember how, as a schoolboy, I took a long while to believe my mathematics teacher when he said that, if red came up on a roulette wheel a number of times in succession, this did not increase the probability that the ball would land on black next time. He finally convinced me by saying: 'Look, the wheel cannot remember what happened before. Each successive throw is just like the first, and there is an equal probability that the ball will land on the red or the black.' I have met many people who are aware of being caught up in this pattern of thinking, but cannot give any logical explanation. It is an interesting question, not easily answered, what actual mechanism it is that one attributes to the roulette wheel. It is almost as though one expected it to get bored with doing the same thing over and over again, and to want to do something else — like a human being.

Far easier to answer is the question why, having repeatedly witnessed a particular sequence of events, we tend to regard the early events as assured preliminaries of those that are to follow. This pattern of thought and behaviour would be absurd if objective reality were a roulette wheel and events happened in random succession. In fact, however, roulette-like successions of random events occur extremely rarely in nature. On the other hand, sequences of phenomena in which transformation of energy produces a regular causal sequence are not merely frequent but observable all around us. When thunder follows lightning, or a downpour follows distant thunder, and even when these events occur in the same order on only a few occasions, there is an over-whelming probability that they stand in a causal relationship to each other. The causation of one event by another always presupposes some form of conversion of energy. And the degree of probability that a particular chain of events is causally linked does in fact increase with the number of times the events occur. What the physicist describes as conservation of energy is no doubt the same objective reality in adaptation to which two independent reaction patterns have evolved. One is the ability to develop conditioned responses, widespread in the animal kingdom; the other is the concept of causality, evolved, so far as we know, only in man.

The avoidance response of the paramecium and the localization in space achieved by the highest vertebrates are both adaptations to another reality — that represented by the first law of physics, stating the impermeability of bodies. Analogously, the abilities of forming conditioned responses and causal thought are adaptations to the law of the conservation of energy.

It is one of the fallacious dogmas of empiricism that the causal thinking of man arises only from habit and that our *propter hoc*, our 'because', is identical with an often experienced, reliable *post hoc*, with 'that which regularly follows'. The axiomatic nature of causal thought nowhere manifests itself more clearly than in the opening sentences of James Prescott Joule's classic paper on the mechanical equivalent of heat. There he states, as naively as dogmatically, that it is obviously absurd to assume that one kind of energy can just disappear without being converted into another. Thus he starts out by postulating what he subsequently proves to be the case, and what, therefore, he had no need to postulate. The *a priori* and axiomatic character of causal thinking finds its equally convincing expression in the insatiable 'why' of intelligent children.

I have made the point several times before that the functioning of simpler cognitive mechanisms can be checked by reference to more complex ones, with the result that the messages from the simpler mechanism are never wrong, but merely provide rather less information. The same relationship exists between conditioned response and causal thinking. Learning by conditioning, like all processes of association, involves selecting, from among the facts embodied in the principle of the transformation of energy, just one — i.e. that cause precedes effect. This is all that is needed in order to enable the organism to prepare for the relevant stimulus situation signalled by the impinging of the conditioned stimulus.

6.7 *Motor learning*

The Viennese zoologist Otto Storch was, I believe, the first to adduce the fact that adaptive modification of behaviour is to be found at a far lower level of development on the receptor side of animal behaviour than on the motor side. With the exception of the elementary process of motor facilitation discussed in the previous chapter, everything that has been said about the teleonomic modification of behaviour by learning refers to receptor processes — sensitation, habituation, acquiring of habits, traumatic association of escape responses, and particularly the increase of selectivity in innate releasing mechanisms, are all processes in which only the receptor apparatus undergoes adaptive changes.

What constitute, then, the most elementary teleonomic modifications of motor functions? The same function of the conditioned response, which by anticipation enables the organism to respond to a conditioned stimulus with appropriate preparations for the unconditioned stimulus that is expected to follow, also enables it to learn the order in which it has to perform particular fixed motor patterns from among its innate stock of such patterns. We are already familiar with an example of this process — the way a rat learns to build its nest. I tend to believe that all motor learning, in so far as it has to do with conditioning and does not consist in mere facilitation, is based on the same principle. In the simpler cases the movements thus strung together in a sequence are self-contained and easily recognizable fixed motor patterns, as in the nest building of the rat. The shorter and simpler the components that are linked together, the more clearly one is aware that a new pattern has been learnt.

A comparatively simple example of such an apparently unified motor pattern is path conditioning in mice. If one observes how a mouse learns to find its way through an elevated maze, one realizes the difference between a free sequence of actions controlled by instantaneous information and a fixed sequence that has been learnt. This can be seen in the films made by O. Koehler and W. Dingler in 1952. When on unfamiliar ground, the mouse advances slowly, step by step, feeling its way with its vibrissae. The third or fourth time it goes through the exercise it may cover a small part of the distance more quickly, then stop suddenly and return to its former method. As it goes through the routine more and more often, it accelerates its movements at more and more points and over ever-increasing distances, until finally these sections all join up with each other. For a long time a slight hesitation at the junctions persists. Its complete disappearance marks the completion of the learning process, and the mouse can run through the whole maze in one go.

The integration of the various segments of fixed locomotor patterns into a unified 'skilled' movement is actually achieved by one conditioned reaction being linked to the next. Each movement is released by the preceding one and creates an anticipated stimulus situation which informs the organism that it is still on the right path.

In his classic work, *Die Orientierung der Tiere im Raum*, Alfred Kühn used the concept of mnemotaxis to describe this type of learning and this kind of orientation mechanism, but it was a mechanism he regarded only as a theoretical possibility. Objections were raised from various quarters that if such assumptions were true the animal would be forced to stay on its one learned path and would be disorientated by deviating from it to the slighest extent. Since Kühn did not know of any animal that met these requirements, he omitted the chapter on mnemotaxis and mnemic homophony in subsequent editions of his work, but this was a mistake, for there are animals, like the water shrew (*Neomys fodiens*), that behave in precisely the way that his theory requires when they lose their way or an experiment is carried out in which the correspondence between their established motor patterns and the physical path conditions is deliberately broken. When I removed from the path of the water shrews a little wooden box they had accustomed themselves to jump on and run across, they jumped into the air at the point where the obstacle used to be, and then sat there on the ground in a state of complete disorientation. Then they began to explore with their vibrissae, and turning back, they recognized a section of the path along which they had come. Gaining confidence, they turned to face forward again, rushed off and jumped into the air again at the very same spot. They reminded me of children who get stuck when reciting a poem and start again a few lines further back in order to try and get over the difficulty — with mnemotaxis!

It seems reasonable to believe that the same principles apply to the learning of more complex motor patterns. Path conditioning consists almost entirely of a linear coupling of components drawn from the action patterns of locomotion, but there seems no reason why simultaneously functioning components should not also be integrated by similar processes.

Some writers have defined the 'learning by heart' of a sequence of motor coordinations as 'kinaesthetic', and, as the word itself suggests (*kinēsis*, movement, and *aisthēsis*, feeling), the feedback from the proprioceptors undoubtedly plays an important part in the acquisition of the motor skill. In phenomenological terms, too, it is an appropriate expression, for one 'feels' one's ability to exercise the motor skill in question. On the other hand, the term suggests that it is proprioceptor memories that make possible the exact repetition of a motor pattern, and this assumption is probably wrong. Erich von Holst showed many years ago that voluntarily produced motor coordinations are also subject to the laws of 'magnetic effect' and central coordination, and this has been confirmed many times since. The researches of John Eccles have shown that the organ that effects the coordination of motor skills is the cerebellum.

As will be further discussed in the section on 'Voluntary Movement', motor learning, even on the highest level, does not differ in principle from the path learning of lower mammals. There are always ready programmed and centrally coordinated patterns innately at the animal's disposal which the learning process merely integrates to form a new unit. As the capacity of animals for motor learning increases with their phyletic development, the motor components become smaller and smaller. But even in the case of genuine voluntary movements these components are far above the level of fibrillar twitching, and no doubt usually include in most cases the contraction of various synergistic muscles; at the same time they are sufficiently small to lend themselves to linkage, both in simultaneous patterns and in linear sequences, and to form an almost infinite number of constellations, or 'motor melodies', as Uexküll called them.

We know with reasonable certainty that the motor elements of locomotion are based on endogenous stimulus production and central coordination. It is my belief that this kind of physiological event cannot be adaptively modified at all by learning — or, indeed, by any other influence. All that can happen is that the multiplicity of processes described by Holst as 'the cloak of reflexes' (see p.58) can be interposed between these physiological events and the demands of objective reality. Support for this view comes from the fact that where the intermediary function of the 'cloak of reflexes' is inadequate, the basic fixed motor pattern does not become 'malleable' and modified by the taxes but breaks down into small units which, by virtue of their brevity, are more easily adaptable to the purposes of spatial exploration.

The survival value of a well-integrated, 'practised' motor skill lies primarily in the fact that it operates quickly. The difference beween a rapidly and purposively executed sequence of movements and a pattern controlled at every step by orientation mechanisms becomes evident when one tries to catch territorial animals living in liberty in their accustomed habitat such as lizards or coral fish, because it is not retarded by a sequence of reaction times. As long as the animal keeps to its familiar path habits, it is too quick and confident for us to hope to catch it in our hand or in a net. But if one succeeds in stampeding it by a sudden movement into an area no longer controlled by its path habits, one can usually catch it. It is, I think, the advantages of these path habits that are largely responsible for causing animals of this ecological type to develop territorial behaviour.

No one has yet attempted to investigate learned motor skills by the methods used by Holst to demonstrate that centrally coordinated actions are independent of afferent processes. To switch off all the afferent nerves, as Holst did, seriously damages the organism, and it would prove little if an animal so treated were no longer able to perform the motor skill learned and practised before. Human beings suffering from *tabes dorsalis*, who are deprived of the proprioceptor afferent nerves informing them about the position of their limbs, suffer from a marked lack of coordination in their movements. One must not forget, however, that man is the world champion in controlled and rationally guided voluntary movements, and that therefore some of his motor patterns may well be susceptible to afferent control while their equivalents in animals are not.

I will not attempt to answer the question as to whether all learned movements in men and animals are physiologically of the same type. I am only concerned with those kinds of behaviour pattern which, as the products of endless practice and repetition, are described in everyday language as 'quite automatic', 'second nature', something one 'could do in one's sleep'. In such cases I do in fact assume that the physiological mechanisms are identical with those on which the path habits of small mammals depend, as described above. To know more about the physiology of these mechanisms would be interesting because the motor patterns in question have many surprising similarities to fixed motor patterns. In the first place, as Holst has shown, certain laws apply equally to the coordination of motor skills and the central coordination of innate motor patterns. Thus, the rhythms of the various basic actions that make up the coordination as a whole influence each other in both cases in the same way. The phenomena of 'relative coordination' and the 'magnet effect', described in Holst's works, produce a harmony between these individual rhythms by bringing them into a phased relationship that can be expressed in terms of low whole numbers. The more perfectly this can be achieved, the more stable is the coordination of the rhythms. Motor patterns that resist this tendency towards central coordination remain unstable — i.e. they are difficult to maintain, as a pianist realizes when he has to play groups of four quavers with one hand and triplets with the other. It is these phenomena of 'relative coordination' and 'magnet effect' that give both the perfectly executed learned motor skill and the unlearned innate pattern the graceful and economical form that appeals so much to our aesthetic sense.

The second characteristic that makes motor skills resemble innate motor patterns is their strong resistance to modification. Karl Bühler used to say that it was an intrinsic feature of what had been learnt that it could be forgotten again — an aphorism which is true, I believe, of learning in the strict sense but not of the process by which coordinations of motor skills arise. For these latter, it seems to me, can never be completely forgotten, and modification and adaptation occur rather through the superimposition of new motor elements over the earlier pattern or through the acquisition of further skills than through the obliteration of a familiar and well-established pattern.

This is the sort of thing one can observe when a driver changes one automobile for another. If there are any actions that can be described as 'what can be done in one's sleep' or 'second nature', they are those of a really good driver. If such a driver drives only one type of automobile over a long period of time and is then required to make a change to another model the inflexibility of these actions becomes very evident. For a long time after my wife changed from an automobile with stick shift to one with steering column shift, she used first to put her hand down towards the non-existent gear stick before moving it up to the steering column. Gradually this movement turned into what anyone not knowing how it had come about would have regarded as a somewhat affected flourish of the hand. When she went back to a car with stick shift traces of this movement still remained. For five years she had objected to the steering column shift and longed to return to the stick shift, yet she could not entirely eliminate the motor coordinations required by column shifting. I leave it to the reader to imagine the elegance of the skilled movements with which the lady now changes gear.

Sports coaches are well acquainted with the way in which movements, once learned, obstinately resist all attempts to get rid of them, and are none too pleased when a young tennis player or swimmer who comes to them to be coached already has some self-taught skill. Far from being an advantage, this acquisition is a handicap to him if he is to attain the optimum degree of skill.

The third and most remarkable parallel between acquired motor skills and fixed motor patterns is that even after lengthy periods of disuse, they both show signs of wanting to be re-activated. One of the strongest forces that lead people to dance, skate or become drawn to some other form of sport is the urge to exercise a particular motor skill, an urge that intensifies with the degree of skill possessed and the degree of difficulty involved.

This same urge to perform difficult motor skills was proved by H. Harlow to exist in macaques, which displayed acquired skills again and again 'for sheer pleasure', not for any reward. The 'joy in the function' (*Funktionslust*, as my teacher Karl Bühler expressed it) clearly plays an important role in everyday human behaviour, by causing us to develop polished, elegant skilled motor patterns, in play as well as work. We know from our own experience how every step in the perfection of a behaviour pattern, every smoothing out of a little unevenness gives us highly pleasurable satisfaction. In my book *Evolution and Modification of Behaviour* I have postulated the existence of a 'perfection-reinforcing mechanism' and voiced some speculations about its effect on the origins of art, the most primeval of which is dancing.

From the phenomenological point of view the mastery of motor skills has various other characteristics, at least one of which is basic to the role of voluntary movements in exploratory behaviour, to be discussed in the following chapter. These characteristics are curiously contradictory. On the one hand motor skills operate to such a degree in our subconscious and unconscious selves that we only hinder them by tracing and trying to check them. Indeed, even in the realms accessible to our consciousness we often do not know what we are doing or how we are doing it. If one has not exercised a particular motor skill for a long while — for instance, if one has not skied for about ten years — one's first thought at the top of a slope is that one cannot possibly do it any more, but the moment one starts to move one is surprised to find that everything works as smoothly as it did before.

On the other hand one often finds that one has so clear a picture of the kinaesthesia of a familiar motor skill that, by observing one's actions, one notices details which have ceased to be consciously self-observed. When my grandson recently asked me which of the two big pedals in my car was the clutch and which the brake, I was surprised to find that I did not know offhand and had to have recourse to this procedure before I could reply.

In my view, such 'kinaesthetic' images of our own actions are of vital importance for the ways in which we apprehend the realities of space, and thus for the faculty of conceptual thought. The movement of the prehensile hand, when steered by spatial insight, and having as one of its most important requisites the tactile feeling of the finger tips — particularly of the right index finger — is, in its turn, the basis on which comprehension and conceptual thought were able to evolve. This view is supported by the extent of the area in the cerebral cortex, both on the sensory and the motor side, in which hand and fingers, particularly the index, are represented, as well as by the connection between the motor areas and the pyramidal tracts, the most important nerve fibres of voluntary movement.

Even in its most elementary form, as in the way the water shrew 'learns by heart' a locomotor pattern, the acquiring of a motor coordination is a cognitive act, and one of remarkable effectiveness, whereas a fixed motor pattern can by its nature only be adapted to such environmental situations as can be expected to confront every individual of that particular species at some time or other.

Skilled movements also display certain functional characteristics found in fixed motor patterns. For example, they are not subject to response delays but possess their own appetitive behaviour and are thus an autonomous motivation. Like fixed motor patterns, they are, so to speak, 'custom built', not only in respect of the general requirements of the species but also of the particular conditions applicable to the individual.

It is easy to appreciate the high survival value of this

combination of characteristics. It is at its highest in animals that have to come to terms with habitats that have a complicated spatial structure and which are variable in an unpredictable way. With tree-dwellers that grip the branches with prehensile hands and feet, and at the same time have to master complicated pathways by acquiring skilled movements, every single movement of hand and foot in the sequence of that motor skill has to be pre-formed. This high degree of precision in adaptive behaviour is especially necessary in such animals because, if the hand is to grip firmly, it has to close round the branch at the right moment and at the right spot. Only slow-moving, nocturnal lemurs like the loris and the potto can afford to leave their orientation to mechanisms that provide instantaneous information. All quick-moving apes and lemurs, on the other hand, especially saltatory species, are masters of skilled movements, and this is one of the causes of man's evolution from a group of such animals.

Since I fail to see that there is any basic difference between the processes involved in the simple acquiring of path habits and those found in the learning of complex motor skills, it does not seem to me necessary to postulate fundamentally different processes in the central nervous system in order to explain receptor and motor learning. The new element introduced by the learning process, on the receptor as on the motor side, consists of new relationships. In other words, adaptive modification probably always affects the synapses and is closely related to what Spemann called induction.

6.8 *Selection pressure and the adaptation of teaching mechanisms*

Through the great 'invention' of the new feedback cycle by which the process of learning by experience comes about, the first as well as the last link in what had hitherto been a linear chain of events acquired a new function. Before the great discovery of the conditioned response was made, the final stage merely served the simple purpose of seeing that a certain 'closed' programme of behaviour ran its course, feeding back the information that this had taken place, and terminating the appetitive behaviour. That this feedback exists can be inferred from the sudden critical waning of the excitation after the consummatory stage has been reached. Proof was provided by F. A. Beach who performed operations on male chimpanzees which had the effect of cutting out the feedback which normally terminates copulatory behaviour — by reporting that the seminal vesicles are empty. The operated animals went on copulating until they were exhausted.

We have already seen (p.58) that all fixed motor patterns tend to run their course continuously without any interruption. The first function in whose service higher centres in the centralized nervous systems have evolved is that of inhibiting the continuous run of centrally coordinated patterns. At the same time these 'centres' have to be able to de-inhibit each of the subordinate patterns at the biologically correct moment; in other words, they have to be capable of receiving instantaneous information on when this moment has been reached. The motor pattern, the inhibiting centre and the releasing mechanism together form that functional unit that Heinroth has termed '*arteigene Triebhandlung*'.

The message passed to the centre by the fixed motor pattern, initially consists only of 'Programme Carried Out', whereupon the centre re-activates the inhibition which has just been removed. Only in exceptional cases is the consummatory act concluded by running its full course, i.e. by exhausting its specific excitation, as with birds whose song gradually dies away.

It is very likely that this simple mechanism provided the opportunities for selection pressure to convert a mechanism that reports the completion of a programme into one that reports its success or failure — a far more demanding function. It is as though a soldier who had only been told to report 'mission accomplished' were suddenly expected to give a reasoned account to the General Staff of the results of his actions and even what had been wrong with his order. For this new function a great deal of exteroceptor and proprioceptor information is necessary, as well as mechanisms with sufficient phylogenetic 'knowledge' at their disposal to be able to distinguish satisfactorily between success and failure.

Thus, the 'invention' of conditioning puts new demands on the mechanism of the consummatory art: not only is it necessary to evolve an altogether new receptor and proprioceptor apparatus checking on the success of the '*Triebhandlung*', the instinctive urge, but in addition much larger quantities of neural energy become necessary to pass a sufficiently 'impressive' report back to the many different neural structures or 'centres' which are concerned with the precedent behaviour that is to be changed by conditioning. Under the selection pressure of these new functions the mechanisms of the consummatory act have developed special structures and physiological characteristics which earlier ethologists failed for a long while to understand, for a non-conditionable system like an instinctive behaviour pattern usually runs its course without any special emphasis on the consummatory act. This is particularly true of patterns which, like the mating of many arthropods, are only performed once in the life of the individual. Here the general level of excitation or arousal does not appear to rise when the consummatory act is carried out. For example, after the excited mating dance performed by the male, the actual copulation, the final link in the chain of wooing the female, appears almost apathetic.

Even with highly developed brood-caring Cichlid fish, which have a marked capacity for learning, oviposition and fertilization take place without any perceptible increase in excitation; in fact, the ritual behaviour of pair formation, with its dance-like movements and its splendid display of brilliant colours, indicates a far more intense arousal than the consummatory act itself. The fish learn a great deal from this; in particular they get to know each other individually, and the permanent attachments that they form depend on this individual knowledge. The conditioning effect of this behaviour is probably greater than that of oviposition and fertilization, and the male fish performs his task with the same dutiful punctiliousness as any other behaviour pattern of parental care. This once led a woman in my research team who was observing the spawning of *Etropus maculatus* to make the unforgettable remark: 'He performs his task with an indifference that borders on virtuousness.'

When one compares mating behaviour of this kind with that of a stallion, or, indeed, with the consummatory acts of higher animals in general, such as the way a predator kills its victim, or a bird of prey strikes, one cannot help sensing in the latter a blaze of excitement that seems to consume the entire organism. This intense general state of excitation is important for the creation of the nervous impulses, sufficient in number and strength, required to influence the innumerable points in the central nervous system where feedback of the consummatory act is to induce modifications. Its intensity is thus not a mere by-product or functionless epiphenomenon but an indispensable part of the physiological mechanism which derives information from success and failure.

In all behavioural systems of the above type the message signifying the conclusion of the action sequence is immediately followed by a critical decrease of tension. But in view of the flare-up of excitation it seems to me rather artificial to regard this decrease as the sole cause of the conditioning effect caused by the performance of the consummatory act. In many other cases Hull's view that relief of tension is the basic conditioning factor, is correct, but it is not so where the consummatory act of the type described is concerned, as one can tell from elementary self-observation.

The releasing mechanism, too, is exposed to a greatly modified selection pressure when in a feedback behaviour system it acquires different functions from those it performed in a non-conditionable linear system. This selection pressure may act in the opposite direction of the one effective in a linear, non-learning system. In the non-variable, linear behavioural chain of an innate motor pattern the innate releasing mechanism is the only neural apparatus that 'knows' when and where the pattern is to be carried out. However, selectivity, hitherto indispensable, becomes not only superfluous but actually disadvantageous when a new process for acquiring information begins to play its role as a teacher. As soon as the feedback from the consummatory act informs the organism of the most favourable time, place and object for performing the pattern in question, it is an advantage for the releasing mechanism to be *less* selective. Consequently we often find in closely related animals very different degrees of selectivity in analogous, and perhaps even homologous, releasing mechanisms. We have already met the unselective mechanism that determines the corvid's choice of nest building material. In the case of weaver finches (*Estrildini*) and weaver birds (*Ploceini*), which have a far lesser capacity for learning, knowledge of the only material that is suitable for their highly differentiated nest building movements is entirely innate, and many of these birds only develop their reproductive urge when the nest material has provided the appropriate key stimuli.

The procedure by which the conditioned response acquires information about an object by the tentative application of an innate motor behaviour pattern has one considerable advantage: namely, that the very characteristics of the object which are essential to the successful execution of the pattern are selectively ascertained. Many higher animals have 'invented' this procedure for conditioning themselves by trial and error to whatever is the most suitable available object on any particular occasion. The bigger the role this procedure plays, the less the innate releasing mechanism need 'know' about the object, and it must in fact aid survival if the releasing mechanism is as simplified as possible and thus gives the greatest possible scope to the learning process. All it then needs to do is to give the young organism a 'gentle hint' as to the direction in which its efforts are likely to be most successful. As already mentioned above (p.58), and as will be explained in detail below (p.144ff.), this form of trial-and-error behaviour differs fundamentally from so-called exploratory or 'curiosity' behaviour.

Chapter seven

The roots of conceptual thought

7.1 *The integration of component functions*

Every one of the cognitive functions to be discussed in this chapter is found in animals, and each has its own survival value, in the service and under the selection pressure of which it evolved independently of all others. If one knew these functions only from animals, it would hardly occur to one that they could be integrated into a system of a higher order. As we have already said (p.33f.), it is as impossible to infer the existence of a higher system from a series of lesser pre-existing systems as it is to deduce the existence of a more advanced animal on the basis of its primitive ancestors. And most of all it is impossible, on the strength of our knowledge of individual component functions, to predict the radically new system characteristics that arise in the new unity created by their integration — the faculty of language and abstract thought, the ability to accumulate supra-individual knowledge, the power to foresee the consequences of one's own actions, and hence the emergence of moral responsibility. In discussing these several cognitive functions I shall be obliged to repeat myself. Many years ago I realized the immense importance which the abstracting functions of perception (particularly of Gestalt perception), the central representation of space, and last but not least exploratory behaviour, each of them found in some variety of animal, must have had for the evolution of man. I have little to add to what I said then. What I had not fully grasped at that time, however, was that the integration of these three cognitive faculties and of at least two others was needed in order to evolve the unique system whose function is conceptual thought, the creation of which constitutes the emergence of man.

The two cognitive processes which I passed over at that time but which must now be discussed here are voluntary movement, which, in conjunction with the feedback that it produces, is a cognitive function *sui generis*, and imitation, which, closely linked with voluntary movement and the reafference of its feedback, provides the basis for the learning of verbal language and therewith the faculty of passing on objective knowledge independently of the presence of its object.

I shall discuss each of these six cognitive functions individually and in the original form in which they are found in animals as though we did not know that they in fact combine as essential components of higher, specifically human functions. Because of their independence of each other it does not greatly matter in which order we take them.

7.2 *The function of abstracting: perception*

Information about external reality that reaches our central nervous system by way of our sense organs never, or only exceptionally, reaches the level of our subjective experience in its original form as separate sense data absorbed by individual receptors. Only in the case of the so-called 'lower' senses does one find that, for instance, a single stimulation of a tactile cell, or a particular scent, travels from the receptor to the mind without undergoing a process of evaluation or interpretation in some way. What our sensory and nervous mechanisms, optical or auditory, convey to us is invariably the product of highly complex if totally unconscious computations which seek to abstract from the chaos of accidental sensory data those data which are constantly inherent in that transubjective reality which, as we hypothetical realists assume, lies behind sense data. The essential function of this 'unconscious reasoning', as has been emphasized (p.13f.), lies in establishing a correlation or constellation of certain stimulus data which remain constant in time, and to do this reliably enough to make these constellations or patterns recognizable. As I have also explained before, every recognition of real things hinges on the procedure that Karl Popper has termed 'pattern matching'. This basic process of cognition consists in matching a configuration prevailing in incoming sensory data with a corresponding pattern existing in our own central nervous system, either as the effect of phylogenetic programming or of individual experience. Spatial configurations which are recognizable in this manner are usually what we call objects. Jacob von Uexküll's simple definition is: an object is that which moves in a body.

Any real thing in our ambient world would be unrecognizable to us if we were dependent on the *absolute* identity of stimulus data and the absolute identity of the configuration in which they appear to us. If, for instance, it were necessary for our visual recognition of an object that we should receive its image always in the same size, form and colour on our retina we should hardly ever recognize it. The wonderful performance of our computing mechanism consists in making our perception and recognition of objects independent of the need for these impossible conditions.

In the sector of vision this computing mechanism has been shown to begin in the retina itself. As has been demonstrated by Lettwin and his co-workers, and later by E. Butenandt, the visual cells on the frog's retina are united into separate groups and send their afferent neurites to one ganglion cell. The latter responds selectively to messages from the group as a whole. One gives a signal when a dark shape passes across the retina from left to right, another when its cells register an increase or a decrease in illumination. There are even ganglia which respond only when a convex area of shadow moves in a particular direction. In the strict physiological sense, the actual stimulus is simply the light that falls on a rod or cone cell. That 'a convex area of shadow is moving across the retina from right to left, is a message transmitted by a highly complex neuro-sensory mechanism which responds to a pattern of individual stimuli.

It is therefore a misleading form of abbreviation when ethologists talk of 'a stimulus', when referring, for example, to the characteristic of 'red underside' to which a sexually active male stickleback responds by going through the motions of attacking a rival. By calling this characteristic a 'key stimulus', we are assuming that its perception depends on information provided by a highly complex computing mechanism which ensures that only this particular stimulus pattern will operate in a certain way.

A great deal of our knowledge rests on the principle of 'pattern matching'. But our perception of patterns involves a process which is the equivalent of abstraction. For if messages from the visual cells in the frog's retina combine to provide information of the kind mentioned above, and if this process functions independently of the absolute size of the stimuli, we are dealing solely with relationships and configurations, i.e. with abstractions.

The ability to perceive a constant relationship between stimulus data irrespective of their qualitative and quantitative variations was identified by the psychologist Christian von Ehrenfels. 'Transposability' is the most important criterion of a Gestalt, nowadays known to science as the Ehrenfels criterion. His classic example is that of a melody that can be recognized at whatever pitch and on whatever instrument it is played. But the ability to pass over what is contingent and to abstract what is essential is not confined to the highly integrated processes of Gestalt perception. As my remarks above on colour constancy (p.11) and the frog's retina prove, the faculty to transpose, to pass over all that is accidental and to abstract that which is essential is characteristic of all perception mechanisms as such and forms the basis of the process of objectivation.

What is abstracted in this way are properties constantly inherent in the object in an invariable form. This can be well demonstrated in elementary functions of perception which we call constancy phenomena. This term denotes a concept determined exclusively by function, for the physiological mechanisms of such different things as colour constancy and form constancy have fundamentally different causal origins, yet all serve the purpose of enabling us to identify the objects around us as being 'the same', even when the stimulus data that impinge on our sense organs are different on each occasion.

Everyone understands what is meant when we talk of the colour of an object, and nobody stops to consider that the object reflects entirely different wavelengths according to lighting conditions. I see the paper in my typewriter as white, although at the moment it is reflecting the yellowish light from an electric bulb; if the red rays of the setting sun were to fall on the paper I would still see it as white. This is effected by my constancy perception which, without my conscious intervention, 'removes' the yellow or red elements from the colour that the paper actually reflects. It excludes from its message the colour of the light — which it has to 'know' for this purpose — since it is normally of no concern to the perceiving organism. Likewise the bee, which possesses a broadly similar colour constancy mechanism, is not in the slightest concerned with the colour of the light but has to be able to identify a rich blossom by its constant and unchanging property of reflecting different wavelengths with different intensities — in other words, by what in common parlance we call its 'colour', irrespective of whether it is the blue light of morning or the red light of evening that is shining on it.

Mechanisms fulfilling similar functions enable us to see the size of an object as one of its invariable properties, although the size of the image it casts on our retina decreases with the square of its distance. Other mechanisms perform the remarkable feat of making the place where we perceive an object appear constant, although the slightest movement of our head, and particularly of our eyes, causes our retinal image of that place to jump to and fro in the most violent manner. The physiology of these two constancy functions has been investigated in particular detail by Erich von Holst, to whose works I would refer the reader at this point.

Far more complex, and as yet hardly studied from the physiological point of view, is the computing mechanism that permits us to perceive the three-dimensional form of an object as constant while it is moving — for instance, turning to and fro — and thus causing considerable variations in the form of the image on our retina. It would require immensely complicated stereometric or descriptive-geometric operations in order to achieve what is for us the natural function of interpreting all these changes in the retinal image, even those of a silhouette, as spatial movements of an object with an unchanging form, not as a change in the form of the object itself.

It sometimes occurs in the evolution of organs, as also in the development of machines, that an apparatus developed to perform one particular function unexpectedly turns out to be able to perform a quite different function as well. It once happened that a calculating machine originally designed to work out compound interest surprised its inventors by showing a capacity to handle integral and differential calculus as well. Something similar is involved with constancy mechanisms of perception, which were developed under the selection pressure of the need to infallibly identify particular objects in the environment. Surprisingly, these same physiological mechanisms are also able to isolate the characteristics not just of one single object but of a whole class of objects, ignoring variable contingent features found only in individual cases and identifying the basic, constant Gestalt of the class.

This supreme function of constancy mechanism is quite independent of rational abstraction. It is equally proper to higher animals as it is to small children. When a one-year-old child calls all dogs 'bow-wow', it is not aware of the taxonomic diagnosis of *Canis familiaris*. Nor, to quote Eibl-Eibesfeldt's example, when his small son called mammals 'bow-wows' and birds 'tweet-tweets', recognizing at the same time that a big goose and a tiny wood-wren belonged to the latter class, and his new-born sister to the former, he had formed the concepts of 'mammal' and 'bird'. In such cases little Adam assigns a name for something which he directly perceives as a generic quality. All these functions of abstraction and objectivation performed by Gestalt perception are not only related to other, simpler functions of constancy perception, but they are actually built upon them, i.e. they include them as indispensable components. Gestalt perception, being a higher integrated system, naturally possesses new system characteristics of its own. Moreover, it is linked to learning and memory.

Perception even appears to possess its own mechanism for storing information. In 'Gestalt Perception as a Source of Scientific Knowledge' (1959), I have described in detail how the process by which a Gestalt or form crystallizes, emerging against a background of contingent elements, may extend over very long periods, sometimes many years. Ethologists and doctors find time and again that a recurrent pattern of individual events, such as a succession of movements or a syndrome of pathological symptoms, is only recognized as an invariable Gestalt after sometimes thousands of observations.

This process is accompanied by remarkable subjective phenomena. Long before one can formulate reasons for it, one often notices that a particular complex of events appears interesting or fascinating. Only after a while does one begin to suspect that there is something regular about them. Both circumstances lead one to repeat one's observations, and the result is then often attributed to 'intuition' or even 'inspiration', although there is nothing supernatural about the process.

What happens in such a case is remarkable enough. We obviously possess a mechanism that is capable of absorbing almost incredible numbers of individual 'observation records', of retaining them over long periods, and on top of all that of evaluating them statistically. One must assume the existence of these two functions in order to explain the undeniable fact that our Gestalt perception is able, presented with a mass of individual images, each of which contains information gathered over long periods of time, and more of it accidental than essential, to work out the underlying invariance. As the jargon of information theory has it: 'Redundancy of information compensates noisiness of channel.'

A system that can achieve this must be highly complex. Yet it is not surprising that, in spite of their many similarities to rational actions, all these sensory and nervous processes take place in areas of our nervous system which are completely inaccessible to our consciousness and our self-observation. Egon Brunswik coined the term 'ratiomorphous' to describe them, thereby indicating that, although they are closely analogous to rational behaviour in both formal and functional respects, they have nothing to do with conscious reason. The suspicion that they might be processes that were initially of a rational nature but had somehow found their way into our subconscious, in the psychoanalytical sense, is easily countered. Highly complex mathematical, stereometric, statistical and other operations have been carried out on ratiomorphous processes in backward children, and even in animals, which even skilled scientists have only been able to copy very imperfectly.

Similarly, the capacity of Gestalt perception to store information for future evaluation is analogous to our rational memory but probably depends on different physiological processes. The ratiomorphous function is vastly superior to the rational function so far as concerns the number of individual items of information it is able to retain, but we are not in a position to call up this information at will. My study of gestalt perception contains further details about the respective abilities and disabilities of these two analogous cognitive mechanisms.

This account of the analogies between rational and ratiomorphous functions is not meant in any way to imply that the conceptualizing, objectivating function of abstract thought has no connection with that of Gestalt perception. Ratiomorphous functions are independent of abstract thought and as old as the hills, for we may assume with certainty that the retina of Stegocephalia from the carboniferous era must have performed the same functions as those performed by the retina of the frog today. From the practical point of view the perceptual functions of objectivation and conceptualization are the precursors of the corresponding functions of abstract thought. However, as is the case whenever pre-existing systems are integrated to form a higher unity, the former are by no means rendered superfluous by the sudden emergence of the latter but constitute its precondition and its component parts.

7.3 *Insight and central representation of space*

The concepts of 'insight-controlled' behaviour and intelligence are closely linked, and generally a creature is called intelligent if it has a highly developed ability to act with insight. Earlier psychologists gave little thought as to how this ability arose, possibly because such a highly intellectual function was considered beyond explanation in physiological terms, and traditional definitions of it were therefore confined to its negative features. Insight applies to a behaviour pattern by means of which an organism comes to control a particular environmental situation in a manner conducive to its survival, despite the fact that it cannot draw on either phylogenetically or individually acquired information about that situation. We have seen in Chapter 1.3 and in further detail in Chapter 4, that we can set against this negative definition a positive definition based on our knowledge of the mechanisms that govern insight-controlled behaviour, namely, the behaviour patterns whose special adaptedness depends on the processes of acquiring short-term information.

The perception of visual depth and direction plays an important part in the orientation of higher vertebrates, and it is not entirely wrong if the naive observer assesses the intelligence of individual species on how far this perception is developed. Animals with binocular vision always seem 'intelligent'; the owl, by no means one of the more intelligent birds, has become the symbol of wisdom. A great number of vertebrates, however, derive their orientation information primarily from parallactic displacements of the retinal image of individual objects caused by the animal's own movements. Spatial perception of this kind is found not only in fish but also in numerous birds and mammals — take the golden plover or the robin, for instance, which locate an object by means of their characteristic bobbing and tilting movements. Ungulates focus only rarely with both eyes, and even a dog running along by its master's side scarcely ever looks at him with both eyes, unless his master calls, sneezes, stumbles or attracts the animal's attention in some such way. Even then the dog usually looks at its master with its head on one side, showing that in the first instance it is orientating itself acoustically.

This parallactic orientation found in so many higher animals has led many people to draw the false conclusion that such animals are not able to focus with both eyes. It must be emphasized that the orientation mechanism of producing the sharpest image of an object by focusing both eyes on it and then ascertaining the distance of the object through the feedback provided by the muscles is a basic faculty of nearly all binocular vertebrates. In the few exceptions to this, such as certain catfish (*Siluridae*) and loach (*Cobitidae*), it is probably a secondary retrograde development. The function which exerted the selection pressure under which binocular vision developed was probably that of the animal focusing both eyes on its quarry.

In many fishes, amphibians and reptiles, the two eyes move independently, as long as the animals are not aiming at their prey. They locate it in the following way: first the animal focuses one eye on its quarry, following its every movement; as the response becomes stronger the other eye is directed to the quarry and becomes fixed on it. Then its head or its whole body is moved to a position in the plane of symmetry between the lines of vision of the two eyes, which continue to focus on the prey. Then the animal edges forward until it is at the right distance for a pounce.

Very probably much the same mechanisms which, as Holst has demonstrated, supply information on the size and distance of a given object to human beings, are functional in most two-eyed vertebrates. The necessary data are provided by reafference from the convergence of the axes of the two eyes and from the close-range focusing of the lenses. Animals with very highly developed mechanisms are not able to catch their prey if it is too close to their mouths. If a seahorse, for example, finds itself in this situation it moves awkwardly backwards, sometimes twisting its body into grotesque shapes so as to achieve the right aim and distance for snapping at the small shrimp, which, swimming towards it, has happened to come too far inside its correct 'range of fire'.

The great majority of vertebrates use parallactic displacement for a rough orientation in space, and binocular focusing for exact localization of their prey. Binocular focusing on inanimate objects, such as the ground surface on which to walk or climb, or anything that might act as an obstacle to locomotion, only occurs when an object of this kind needs to be located with precision, for instance when an animal needs to step on it or grasp it securely in the course of locomotion. Some fish also have to cope with this situation, such as certain bottom-living or climbing forms which, being heavier than water, have lost their faculty for effortless floating. They stay on the bottom, except on the occasions when by dint of strenuous swimming movements, they manage to lift themselves off it. If such creatures are to move purposefully in such uneven and varied surroundings as a rocky coast or a tangle of mangrove roots, they need to be able to locate with precision the details of the ground over which they crawl or swim.

Several species of perch-like fish (*Percomorphae*) have evolved special features for this purpose. Especially well equipped are certain blennies (*Blenniidae*) and gobies (*Gobiidae*). Almost every observer of these interesting little fish has drawn attention to their 'intelligence'. William Beebe, for example, said, 'Of all fishes they are the least bound up in fishiness'; while my teacher Heinrich Josef made the jocular but pertinent comment: 'The blenny is not really a fish at all but a dachshund.' This is a good example of the amusement felt by the experienced observer whenever an animal does something unexpected, especially something only associated with more advanced creatures. When one watches a blenny hopping 'on foot' as it were right up to a big stone, rear itself up on its pelvic fins, which are placed very far forward, then raise its head sharply upwards (in almost all other fish the head is rigidly attached to the body) and focusing both eyes on the top of the stone before finally making its precisely measured jump, the effect is irresistibly comical. Even more entertaining and accomplished is the walking fish, or jumping fish (*Periophthalmus*), which also belongs to the family of *Gobiidae*. It surpasses all other members of this group in the ability to focus on inanimate objects, and also in climbing. It clambers up mangrove roots and can jump from one to another with remarkable precision of aim.

Among both blennies and gobies there are primitive forms which are not differentiated in the way just described but whose fully functional air bladder enables them, like other fish, to move about freely in the water without muscular effort. If one compares these with the climbing types belonging to their respective order one is struck by a basic difference in the shape of their heads. The former, like other fish, have a flat forehead, with the eyes on each side of the head; but the greater the adaptation to living on the bottom and to climbing, and the more important, therefore, the function of precise optical orientation, the more angular the forehead becomes and the closer the eyes move to the ridge that separates the front of the head from the dorsal line, so that the frontal view for binocular focusing is unobstructed. This adaptation is at its most fully developed in *Periophthalmus*. Fig. 2 shows the various stages of development.

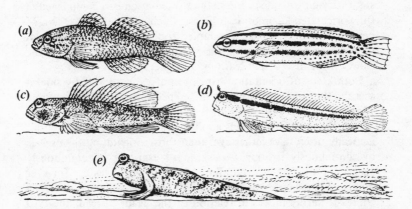

Figure 2 Convergent adaptation of the position of the eyes to binocular focusing in gobies and blennies. The head of the free-swimming *Gobius Dormitator* (a) and of the *Blennius petroscirtes* (b) has the same shape as that of most free-swimming fish. Bottom-living fish, like *Gobius jozo* (c) and *Blennius rouxi* (d) have a steeply profiled head which enables them to focus sharply on objects ahead of them. *Periophthalmus* (e), which lives on land, shows the adaptation at its most advanced.

Exploitation of parallactic displacement and location by binocular focusing are two different but equivalent mechanisms. Nevertheless, the anthropomorphic view that focusing animals are more intelligent than those with fixed vision is not entirely false. Fish most familiar to us, like goldfish, have the proverbial 'fish-like stare'; they can only receive orientation information as long as they move — that is, as long as the different retinal images of objects keep shifting their relative positions. When a fish of this kind has stopped in front of an obstacle, such as a stone or a thick clump of vegetation, it often first resumes its motion on a 'collision course', e.g. in the very direction that is blocked. Only when this movement has produced a corresponding displacement of the retinal images and supplied the appropriate orientation information does the fish start to change its course.

The behaviour of animals that orientate themselves by focusing differs from this in one fundamental respect: that whereas in parallactic orientation the actions of locomotion and location coincide, in the case of focusing animals location precedes locomotion. Animals that orientate themselves telotactically, i.e. by bringing one object after another into sharp focus and thereby getting to know their immediate surroundings, invariably remain stationary during this process. Once orientation is accomplished, a fully orientated locomotion follows, which gives the impression of being planned. We shall encounter similar patterns of internal preparation followed by motoric execution when we discuss the functions of insight and learning at higher levels.

Both kinds of orientation can, of course, occur side by side in one and the same organism, the difference in their role being purely quantitative. Peripheral vision is important for parallactic orientation; consequently animals good at orientating themselves parallactically often have larger eyes than those orientating themselves by continuously fixating. A siskin, for example, is continuously focusing on objects surrounding it, and its tiny eyes are for ever moving. The large eye of the robin or the golden plover, by contrast, hardly moves in relation to the head; instead, the bird carries out, in regular sequence, the bobbing movements mentioned above.

Strangely enough, in many cases the two methods of orientation are not interchangeable. As I have said above, the primary purpose of binocular focusing is to capture prey, while parallactic perception provides approximate orientation and enables the animal to avoid obstacles. Some deep water fish which have the excellent binocular vision of a predator are not able to use it in order to avoid obstacles. On the Adriatic coast I once saw hundreds of young fish — a species of garfish (*Belonidae*) — simply swim to the shore and die. They did not come in a swarm but one by one, about a yard apart but strictly parallel to each other, apparently guided by light, salinity or the like. A few yards from the shore they were still in good shape; in the surf they fought to survive; on the beach their dead bodies lay piled up in a row. We know from deep-water fish kept in captivity that some of them are unable to cope with obstructions which are at right angles to the direction in which they are swimming, and even in large tanks they frequently knock their heads against the side. In the open sea their parallactic orientation causes them to swim in a wide arc round any large, opaque object they see from a distance, and they cannot cope with a situation in which they are surrounded by such objects.

Our discussion above of the adaptation of blennies and gobies to climbing made us aware of the close correlation between the faculty of spatial orientation and the structure of the habitat. This is a universal correlation in the animal kingdom. Only a minimum degree of such ability is required from organisms that live in the spatially homogeneous medium of the deep sea. The 'stupidest' free-moving multicellular animals at present known to us are probably various jellyfish which, like *Rhizostoma pulmo* are completely without orientation responses. They have no need of such responses, however, because their lightbell and heavy stalk or column preserve their equilibrium, and because they filter their nourishment from the water without having to catch individual particles of food. The only orientated movements found in these creatures are those of the amoeboid cells of the stomach wall, and the only response the organism as a whole makes to stimuli is that one contraction of the bell stimulates receptors which cause an adjacent contraction. 'It perceives nothing except the tolling bell', as Uexküll wittily put it. The whole beast is more stupid than its single amoeboid cells.

We have already observed the inability of certain deep-water fish to react purposefully to spatial structure. But a close connection between the nature of the habitat and the ability to come to terms with it by 'intelligent' orientation behaviour is also found in land animals. One might regard prairie country, for example, as a kind of terrestrial equivalent of the open sea. If one compares the orientation ability of closely related animals, one of them a prairie-dweller, the other a rock- or tree-dweller (partridge and quail, say, or oryx and chamois), one discovers surprising differences in their ability to 'understand' spatial structures.

Thus a pair of partridges reared from the egg proved unable to recognize a whitewashed wall as an impenetrable obstruction. When they were released, they ran towards the wall opposite the window, because it was brighter than the wall underneath the window. Having reached it, they did not just touch it but pressed against it with their beaks and ran ceaselessly along it trying to penetrate it. Each time they turned at the end of the wall, they pressed forward so hard that they gradually rubbed away the horny layer on their upper beak and the feathers at the front of their neck, which forced me to cover the wall with a dark curtain to stop them doing any more harm to themselves. In order to confine the birds to one part of the room, I put up a board that was just high enough to prevent them looking over the top. They often ran up and down in front of the board but never learned to fly straight over it, even when they kept on finding themselves on the other side of it after having aimlessly taken wing.

The birds were strikingly better orientated in flight than on the ground. When they suddenly took off with the powerful acceleration characteristic of them, I expected to hear a loud crash as they struck the wall or the window, and then to find a dead bird lying on the floor. But nothing like this ever happened. As soon as they approached a wall, they would invariably turn away and settle on the floor. It is obvious that while running on the ground the partridge can 'afford' to pay no attention to obstacles, whereas in flight it has to be able to avoid vertical obstructions like trees and hedges, the edge of a forest and precipitous hillsides. When it is walking on the ground it can no more use the power it has to orientate than the garfish can use the binocular vision it has for catching its prey in order to avoid colliding with a steep cliff.

As I have said, the faculties for coping with complex spatial parameters can differ considerably in closely related types. The garfish, which is not even capable of 'understanding' that a steeply shelving beach is an obstruction, has freshwater relatives which are able to adjust to the glass sides of a small aquarium, even though, as one can see from many photographs, they usually injure their pointed jaws in the process. The Californian quail, for instance, which is closely related to the partridge, has highly developed orientation mechanisms and there are many other examples. Since one can hardly assume that the sense organs of these animals are differently organized from those of their 'stupider' relatives, one must assume that the variations in orientational skill derive from the central nervous system.

In what habitat, one may ask, are animals required to develop most fully their faculties of spatial orientation? The answer is: in trees. And of animals that live in trees, the ones that need the most precise and most detailed information about their spatial parameters are not those that hold on to the branches with suctorial pads or claws, like tree frogs or squirrels, but those that climb by means of grasping extremities which grip the branches like pincers, such as chameleons, some arboreal marsupials and lemurs. The former can risk projecting themselves in the approximate direction they want to go, since just one or two toes are enough to hold on by; even if they were to fall nothing too serious need happen, since they are usually quite small. A grasping extremity, however, can support the body only when it closes round the object in the proper situation, at the proper spot, and at the proper moment; it cannot grip if it stays either open or clenched like a fist.

There is a correlation in mammals between their use of grasping extremities and the position of the eyes in their head, as there is in fish between the eye position and reduction of the air bladder causing inability to float, and for similar reasons. In both cases it is caused by the necessity of very exact spatial orientation. Among both marsupials and placentals there are tree-dwellers of the claw type and of the grasping type, the former generally having their eyes at the sides and somewhat protruding, like the squirrel, the others usually having eyes that look forwards, like those of an owl. With the marsupials the correlation is less evident, probably because there are many very slow climbing kinds whose variety of modes of locomotion makes it unnecessary for them to possess a particularly well-developed orientational ability. It is clear, however, that all animals which leap great distances and then hold fast to their objective with one hand, do indeed focus on this objective beforehand with both eyes.

The hooked hand with which anthropoids climb and swing from branch to branch demands no less a precision of orientation than does the grasping hand of other primates. But there is a further peculiarity of the anthropoïd's unciform hand to which I shall return in the chapter on curiosity and self-exploration — namely, that it operates in the immediate vicinity of the focusing point at the moment it grasps the object.

In describing the behaviour of animals that orientate themselves by focusing on objects around them, I said that the fact that they first explore their environment with their eyes, then make the appropriate movement, gives the impression of remarkable 'intelligence', because the process is analogous to the constructive thinking that we find in anthropoid apes. When an anthropoid is set a problem which can be solved by insight, it behaves quite differently from a raccoon or a rhesus monkey faced with the same situation. The latter keep running up and down, actually trying one potential solution after another. The anthropoid sits down quietly, reviews the situation most carefully, gazing attentively from one part of the experimental set-up to the other. Its inner tension becomes apparent only by its frequent gestures of 'embarrassment', such as scratching its head like a man in thought, or other 'displacement' activities.

A film of experiments made with an orang-utan in Suchum, in the Soviet Union, shows this very clearly. The animal was given the task of pushing a box from one corner of the room to the diagonally opposite corner, where there was a banana hanging from the ceiling by a piece of string. To start with, the orang-utan looked helplessly up and down from the box standing in one corner to the banana hanging in the other; then, in a fit of bad temper, it tried to turn its back on the problem — what M.R.A. Chance called a 'cut-out' action. But this it found itself unable to do, and it turned its mind to the task again. Then suddenly its eyes moved from the box to the point on the floor immediately underneath the banana, from the floor upwards to the banana itself, then down again and from that spot back to the box. In a flash, as one can clearly see from the orang's expressive face, it realizes the answer: turning head-over-heels with delight, it immediately goes over to the box, pushes it underneath the banana and claims its reward. Once it has found the solution the animal takes little more than a few seconds to do what is required. No one who has watched an ape solve that kind of problem can seriously doubt that at the moment it finds the solution, the animal has a flash of insight like that experienced by human beings in such situations — the 'Aha experience', as Karl Böhler called it.

What, objectively and subjectively, goes on inside the orang's head while sitting, externally quiescent, it performs hard mental labour? What the ape actually experiences we cannot tell; but we can say with virtual certainty that the process as a whole is analogous to that which we call thought. It is my own view that the animal does just what I do — it performs imagined acts in an imaginary space with models of spatial objects represented in its central nervous system. It finds the solution by pushing the central representation of a box around in centrally represented space.

I cannot see that, basically, thinking can be other than tentative exploratory actions of this kind taking place in a neural 'model' of outer reality. At the very least I would maintain that such processes take part in and are at the foundation of our highest mental activities. I cannot conceive of any form of thinking that is independent of this foundation. Over twenty years ago W. Porzig wrote the following words in his book *Das Wunder der Sprache*:

> Language translates all non-visual relationships into spatial relationships. All languages do this, without exception, not just one or a group. This is one of the invariable characteristics of human language. Thus time relationships are expressed as space relationships — 'before' Christmas, or 'within' two years. In the realm of psychology we talk not only of internal and external, but also of things being 'above' or 'below' the threshold of consciousness, of the 'sub'-conscious, of 'foreground' and 'background', of the 'depths' of soul. Then we say: ' "Besides" his work, he gives private tuition', or 'His love was "greater" than his ambition', or ' "Behind" these measures lies the intention ...', and so on. The importance of this phenomenon rests on its universality and on the vital part that it has played in the history of language. It can be demonstrated not only from prepositions, all of which originally had connotations of space, but also from verbs and adjectives.

In 1954 I wrote in 'Psychology and Phylogeny' that these views were 'of fundamental importance not only for the history of language but even more for the entire phylogenetic development of thought, including pre-verbal and non-verbal thought'. The recent work of Noam Chomsky and his collaborators has provided strong confirmation of this view.[4] On the basis of comparative linguistic studies Chomsky came to the conclusion that certain fundamental structures in language and thought are innate and to be found in identical form in human beings of all civilizations. According to Chomsky, this is the product of the selection pressure, not of communication, but of logical thought. Independently of Chomsky, Gerhard Höpp arrived at a similar conclusion in his work *Evolution der Sprache und Vernunft* (1970): 'Language is not only a means of communciations but an integral part of reason itself.'

7.4 *Insight and learning*

Earlier (p. 25ff.), insight-controlled behaviour was defined as a function of the mechanisms of instantaneous information. The highest of these functions, however, also involve learning processes, as many learning processes, in their turn, also include elements of insight behaviour.

In all relatively complex processes of insight-controlled behaviour there are mechanisms at work which acquire instantaneous information in chronological sequence. The earliest of this information has somehow to be stored, so that it can subsequently be combined with later information to produce meaningful behaviour. We have also seen that anthropoids *never* act at once when faced with a particular problem that can be solved by insight. Even at a lower level the oriented movement as we have seen follows a period of quiescence, during which the animal looks around in all directions and collects information on the spatial situation.

In such a case, and even more so in insight-controlled behaviour of a more complex nature, the items of instantaneous information provided by the appropriate mechanisms have to be compared and correlated, and it is the function of a system created by the synthesis of these mechanisms to find a solution to the problem in question. A conscious effort of memory must also be involved. In his classic experiments on the insight-controlled behaviour of chimpanzees, Wolfgang Köhler describes how the animals surveyed the individual aspects of their tasks systematically, one by one, and 'committed them to memory' as it were. The orang-utan referred to above behaved in a similar manner.

Another combination of the functions of insight and learning is found in the most elementary forms of learning by trial and error. Even where, as described on p.55f, a single fixed motor pattern is tried out on various objects, actual 'exploratory' behaviour not yet being in evidence, the initial sallies of the still quite inexperienced animal are never entirely haphazard, like the 'experiments' of the genome. And even where the animal does not appear to possess the slightest information that could help to solve the particular problem — if, for instance, one shuts up a cat in a puzzle box — its attempts at solution always begin in a direction specified by instantaneous information. Nor do the animal's efforts simply amount to haphazard combinations of muscular contraction: the cat does not try to get out of the box by closing one of its eyes or licking one of its paws but embarks at once on more 'sensible' courses of action such as scratching at the sides of the box or trying to poke its nose and paws into chinks or corners — in fact, by directing all its efforts from the very beginning towards the points which seem to offer the best chance of escape. Its actions are based on a large stock of instantaneous information which add considerably to its chance of success and, from the functional point of view, are analogous to a working hypothesis.

In sum: there are no functions of more advanced insight-controlled behaviour that do not depend to some degree on learning and memory. On the other hand, the process of learning by trial and error is always subject to regulation by orientation mechanisms which are inseparable from insight-controlled behaviour. Elements of such behaviour also play a considerable part in the process of learning by curiosity, as we shall see later.

Finally I wish to discuss in this section one further relationship between learning and insight-controlled behaviour which Köhler emphasized in his experiments with chimpanzees, a relationship which shows the two in conflict with each other. A sequence of actions which, when first performed, was clearly the product of insight becomes consolidated, after being repeated several times, into a regular learned pattern. If one now makes a slight alteration in the problem one sets the animal, which does not actually make it more difficult but does mean that the routine for solving it in its original form no longer meets the situation, the animal cannot cope, simply because it cannot break its behavioural routine. In fact, the smaller the change required, the more completely the force of habit prevents a solution which an unprejudiced animal would find without difficulty.

Norman Maier studied the same phenomenon in rats. He also carried out the following amusing experiment with human beings. Having assembled a sizeable number of people in a gymnasium, he told them to tie together two ropes that were hanging from the ceiling, the ropes being too far apart for anybody to hold one of them in one hand and reach the other with the other hand. The only aid available was a large stone, and the solution lay in tying one rope round this stone, making it swing and then, holding the other rope at the same time, catching the rope with the stone when it reached the furthest point of its swing. A remarkably small percentage of those present, only slightly more than 60 per cent, found the solution. Maier then set another group of people the same problem, but this time substituting a long hook for a stone. The hook was much easier to attach to the rope and just as effective as a weight, yet the percentage of people who solved the problem fell to a little over 50 per cent. The explanation lay in the fact that a considerable number remained preoccupied with a fruitless attempt to use the hook as a hook, that is to say, to try and hold one rope in one hand and reach the other rope with the hook, which the distance between the ropes had purposely rendered impossible.

Maier drew the incontrovertible conclusion that although all insight-controlled behaviour in problem situations depends on what has already been learned — no one could have solved the problem in question unless he had had experience of various pendulum movements — the actual finding of a solution is easily prevented by adherence to habits of thought and ways in which one has learned to do things. Maier defined the ability to solve such problems by insight as the readiness to fundamentally change one's approach.

The history of science offers many examples of how routines and habits of thought have obstinately prevented the discovery of solutions for problems that are far from difficult. This inhibiting influence also explains why great discoveries are so often made by people who are not experts in the field in question. The second law of thermodynamics was not discovered by a physicist but by a doctor, Robert Julius Maier; the syphilis virus was traced neither by a bacteriologist nor a pathologist but by Schaudinn, a zoologist. Schaudinn himself told my father how it came about: doing simply what any zoologist would do as the first stage in microscopic examination, he looked at the syphilitic secretion in its unfixed, uncoloured, 'natural' state and immediately noticed a mass of spirochaetes. 'Are there always that many spirochaetes in a smear from a syphilitic ulcer?' he asked the pathologist friend in whose laboratory they happened to be. 'Certainly not', came the astonished reply. For spirochaetes cannot be coloured, and are therefore invisible in any preparation produced by the usual methods of bacteriology and pathological histology.

7.5 *Voluntary movement*

It is understandable that the evolution of orientating mechanisms, and therewith of an ever increasing ability to form precise and detailed images of space, should be accompanied by a corresponding increase in the range and subtlety of movements. Without this the organism would be unable to take account in its behaviour of all the details of the abundant information. That is to say, without a corresponding evolution in motor behaviour, the development of powers of orientation and insight would be futile. This does not mean that the development of motor behaviour adjusted itself to that of orientation, for both arose in response to the demands of a complex environment, and must therefore have emerged hand in hand with each other. Nevertheless, their physiological independence can be clearly seen from the rare cases in which the one precedes the other, and I shall now discuss a few examples of this.

There is any conceivable number of intermediate stages between centrally coordinated motor patterns which are quite inaccessible to the control of orientation mechanisms and the almost infinitely variable and adaptable patterns at the disposal of insight. A good example of the former is the courtship behaviour of certain anatids, while the latter is illustrated by the voluntary movements discussed earlier (p.103).

A number of earlier scientists combined the functions both of orientating and of locomotor mechanisms under the single concept of taxis, which, from the functional point of view, is accurate enough; and my earlier accounts of positive or negative phototaxis or heliotaxis, like those of positive or negative tropisms, referred not only to changes of direction but also to movement in this new direction. Our concern here, however, is with the relationship between the cognitive functions of spatial orientation and the motor processes they control, so that we must keep the two apart, distinguishing, to use a familiar image, between the activity of the captain who draws on various sources of information in order to calculate the position of his ship and decide on the best course to steer, and the motor functions of the vessel that carries out his decisions.

The nature and scope of a captain's power over his ship is analogous to the influence exerted on the motor behaviour of an animal by orientation and insight. In the first place the ship's engines, like the animal's motor behaviour, can be 'told' to stop or start, and also in some cases to go into reverse; in the second place superimposed on locomotion are the changes of direction, controlled respectively by movements of the helm and various topic responses (see p.51f.) and motivated by environmental conditions. These two possibilities can be realized simultaneously or successively, in a variety of forms.

A simple stop-go control is probably only to be found in the most primitive of unicellular organisms which seek the most favourable environment in accordance with the principle of kinesis (see p.49). Many such organisms — certain flagellates, for instance — do not appear even to have the power to 'tell' their locomotor mechanism to stop but can only 'decide' between slow and full speed ahead. No one seems to know whether there are flagellates that can also go into reverse.

Ciliated infusoria like the paramecium can, as we have already seen, do more than this. They have control over at least three different regions of their body surface, which is covered with rowing cilia, and in each region they can make the cilia drive forwards or backwards, or stop, independently of other regions. This faculty is the basis of their phobic and topic responses. The practical application of the principle of the combination of mechanisms that can each be stopped and started at will has been known to technology for a long while. Every lover of Mark Twain knows that each paddle of the old Mississippi steamboats was driven by a separate engine. Modern caterpillar tractors are another example. These would appear to be the only realization that Jacques Loeb's principle of tropism has found in the organic world.

The second way of making an animal move in a particular direction is by means of an apparatus independent of the driving mechanism, such as the rudder in a ship, or the lateral curvature of the body in many vertebrates. There are examples of this even on the level of unicellular organisms. The flagellate *Euglena*, for instance, which contains chlorophyll, finds it necessary to live in zones where the level of illumination is sufficient for photosynthesis to take place. It reaches these zones by first turning in a graceful curve to face the source of light, then making straight for it. Metzner has demonstrated how this orientation comes about. Like many unicellular organisms, *Euglena* swims forwards while rotating on its body axis. Its orientation is achieved by two organelles, a light-sensitive area in the protoplasm and a bright red stigma (eye spot). When, with the rotation, the stigma casts a shadow on the sensitive area, this elicits a sudden beat of the flagellum, which causes the front end of the organism to turn a constant number of degrees towards the direction of the light. These little movements are repeated until the creature finally swims straight towards the light, when the stigma circles the sensitive area without throwing a shadow on it. Doubt was cast on Metzner's conclusions at the time, but they are supported by the exact circularity of the path the creature describes.

An example of behaviour in which a fixed motor pattern in a higher vertebrate is controlled by a taxis, as the ship's movements are controlled by the rudder, was investigated by Tinbergen and myself in 1938. When a greylag goose rolls a stray egg back into the nest cup, the action it performs with its head and neck through the sagittal plane is absolutely invariable, beyond the influence of any exterior stimuli both with regard to its form and the strength expended on it. The motor pattern simply 'jams' if the bird is given an object to roll which is too big, and fails from weakness if it is only slightly heavier than a goose's egg; if it is lighter it lifts it off the ground. In order to keep the egg balanced on the end of the beak throughout the operation, the goose makes small lateral balancing movements of head and beak in response to tactile stimuli felt on the underside of the beak. If, with a dexterous snatch, one suddenly takes the egg away after the bird has started, it continues the egg-rolling movement to the end, without the egg, but also without the balancing movements. These movements also stop if one gives the goose a wooden cube to 'roll' instead of an egg: the cube rests safely across the jaws on the lower part of the bird's beak, and there is no danger of it rolling off, as there is with an egg. People who have seen the film we made of this exercise are always surprised at how mechanical the process is, seeing that in other respects the goose is so intelligent a creature.

We have thus had examples of the two principles which enable the higher centres of the organism to exercise effective control over its motor behaviour — firstly the division of the fixed motor pattern into appropriate segments by means of inhibiting and releasing, or stop-go, and secondly the steering of a motor coordination by a superimposed simultaneous movement dependent on external stimuli. There are very few cases which are dominated by the one or the other process to the extent revealed by my chosen examples, for generally, especially in higher animals, both are present side by side and combine in numerous ways.

The processes of inhibition and de-inhibition or release retain an important role even in the most advanced animals. It can be shown that not only in earthworms but also in humans endogenous-automatic, centrally coordinated behaviour patterns would repeat themselves *ad infinitum* if they were not centrally inhibited when not required. In cases of brain damage, such as often result from encephalitis, some fixed motor patterns run on and on, such as sucking movements with the mouth or clenching movements with the hands.

Even in lower vertebrates the 'highest centres' within the central nervous system have other significant functions besides the inhibition and de-inhibition of fixed motor patterns. The cerebral ganglion, the worm's 'brain', decides not only whether, and to what extent, such patterns shall be inhibited and de-inhibited but also which of the various available patterns is to be employed at any one time.

To be sure, a fixed motor pattern with its inhibition mechanisms is a self-contained and comparatively independent system, but very often such a mechanism, which only serves one particular function, does not always require or involve the animal's entire muscular system, so that there is also scope for other motor processes. As in the egg rolling of the greylag goose, inherited motor patterns and taxes are simultaneously in evidence, so we can also find one motor coordination superimposed on another in a single behaviour pattern. This can take place on various levels of integration. We find in fish, for example, that two locomotive organs, such as pectoral fins and caudal fin, can respond to two opposing motor impulses at the same time, the former propelling the fish forward, the latter backwards. As Holst showed in experiments on reflex and endogenous-automatic fin movements in wrasses (*Labridae*) antagonistic impulses may also be superimposed in a single muscular contraction.

It was said at the beginning of this chapter that the demands made on the adaptability of an animal's motor patterns increase with the growing complexity of the cognitive processes that provide insight into complex structures of the environment. Both the principles mentioned above are needed to meet these demands, and it is instructive to investigate to what extent closely related animals from different environments are able to control their own motor activity. The less homogeneous the environment, the smaller the minimum indivisible fixed motor pattern has to be. With quick-moving animals that live on the open prairie, the influence of orientation mechanisms on locomotor patterns goes only slightly beyond the control exerted by the captain over the engine of his ship; that is to say, the animal may choose to walk, trot or gallop, but only as self-contained coordinated movements, so-called gaits. New combinations, as a horseman knows, can only be achieved by patient training. Zeeb and Trumler have shown that all the movements which the horses of the Spanish Riding School in Vienna have been trained to perform are fixed motor patterns in each horse; it is only the release of these patterns by the horseman's commands that depends on the development of conditioned responses.

With an animal that lives in open country the smallest component of the motor coordination may be comparatively large, for the terrain over which the animal decides to gallop does not vary very much from one gallop to the next. If a change does occur, the obstruction can usually be recognized far enough away for the animal to either stop or take avoiding action. But, as we know, horses can very easily stumble over unexpected obstacles.

The control exerted by orientation responses over a horse's locomotor coordination is extremely slight. Riding a horse up an uneven trail on a mountainside, one notices that although it does not just amble blindly on but watches the trail and tries to put its feet in places that provide a firm foothold, the aim-taking process is very approximate. It cannot aim to plant its feet on a definite spot; it cannot, for example, step from one boulder to another. Donkeys and mules are far more skilled in this respect, while mountain zebras are said to be masters of the technique.

A situation analogous to that of the equids of plain and mountain is found among antelopes. The species living on steppes behave rather like the horse, whereas the mountain antelope, the chamois, is more surefooted and adaptable in its movements than any other mammal except primates. The amazing thing is that the chamois has the ability to choose the precise spot where to set its feet, while maintaining the economical coordination of its gallop. Even when it is moving over ground covered with boulders of different sizes the flowing rhythm of its gallop retains its harmony. Only now and then a little syncope or hesitation, which actually makes the movement even more elegant, hints that the taxes are intercalated, and that occasionally the animal finds itself forced to resort to the method of inhibiting and de-inhibiting in order to adapt its movement to the irregular structure of the surface.

The need to break down fixed coordination patterns in the interests of adaptation to the spatial structures emerges with special clarity when, as mentioned above (p.133), an animal's knowledge of the environment would be sufficient for it to deal with a particular problem but a solution is frustrated by the inadequacy of the available motor pattern. There is a proverbial joke among the ferrymen on the Danube which provides a graphic analogy to this comparatively rare occurrence. The captain of a paddle steamer who had made a mess of coming alongside the quay and was trying to do so by manoeuvering his boat backwards and forwards, shouted to the engine room, 'Three turns ahead!', then 'Four turns astern!' then 'Five turns ahead!', and so on, all to no avail. Finally he shouted in desperation, 'Two revs sideways!' If the boat had been equipped with a Voith-Schneider propeller, it could have carried out the captain's order.

There is a parallel to this story in the behaviour of a greylag goose which had learnt to climb up and down a flight of stairs — the latter process being by far the more difficult. If the treads are broader and higher than the goose's stride, it is incapable of correcting the ever-increasing discrepancy of phase by interposing an extra step. Eventually one foot lands so near the back of a tread that when the bird makes its next step, the back of its leg grazes the front edge of the same tread and its foot cannot reach the one below. It pulls its foot back but then tries over and over again, without success, to plant it on the next tread. Finally it makes use of its wings, leaves the one leg dangling and hops down to the next tread on the other. Its actions are now in phase again, and it descends a few more stairs until the same situation is reached again, whereupon the process is repeated.

Muscovy ducks and Carolina wood ducks have no greater orientational insight than greylag geese, but being true tree-dwellers, they have a motor ability that the greylag goose lacks, i.e. if they find that the next stair is too far below, they first make a calculated short step to the front edge of the stair on which they are already standing.

There is a further situation in which the greylag goose is seen to be forced 'against its better judgement' to indulge in quite pointless behaviour as a result of its inadequate control over its motor activity. If it finds itself in the position of having to climb over an obstacle which is about breast height, such as a wooden border round a strip of lawn, one can observe how, while it is still several yards away from the obstacle, it already realizes what it is going to have to do. It begins to lift its feet higher and higher, and a whole step before reaching the barrier it often raises its foot higher than the obstacle itself. In such a case its foot rarely comes to stand fair and square on the top edge of the barrier; if it steps too far over it — which happens just as frequently as stepping short — the bird calls its wings to its aid, as described above. In exceptional cases the goose has been known to move quietly up to the barrier, instead of approaching it with its ludicrous 'goose-step', focus on the top surface, its neck drawn back quivering with excitement, jump up on it with both feet and down immediately on the other side. Muscovy ducks and Carolina wood ducks always behave in this way, without, however, displaying any particular signs of agitation.

The 'invention' made by evolution of cutting segments out of an extensive inherited motor sequence so that it becomes an independent behavioural element, at the disposal of insight, probably constituted the first step towards the emergence of what we call voluntary movement. With the latter the cut-out element has one essential characteristic in common — namely, it can be combined with others of its kind to form a new pattern which is adapted to specific external circumstances and which, like an inherited pattern, runs its course without interruption and without delay by reaction times. In phylogenetic terms, this 'acquired motor activity' appears later than 'acquired receptor processes' (the terms are Otto Storch's: see p.101). It has already been pointed out that the most elementary form of motor learning is path conditioning, and that more complex behaviour patterns are probably also learnt in the same way.

We know that endogenous stimulus-production and central coordination form the foundation of locomotor behaviour patterns. The impulse sequences of these patterns do not appear to be at all susceptible to modification by learning, and their apparent 'flexibility' derives from the intercalation of a profusion of processes between them and the external world — Holst's 'cloak of reflexes' again. As far as I can see, the intermediary function of this mechanism rests on the two processes I have described: either an orientation response is superimposed upon the basic fixed motor pattern, or if this pattern is too long or too rigid, it is cut up into sections which are short enough to be linked together in new combinations fulfilling all the demands of spatial insight.

What are traditionally called 'voluntary movements' in human beings are usually products of motor learning — that is, 'skilled movements' composed of short motor elements. The smallest of these, as has already been said (p.103), are invariably on a level of integration far higher than that of fibrillar contraction. Strictly speaking, the term voluntary movement should apply to these individual micro-patterns *before* they have been combined by a learning process to form a complete smoothly running skilled movement. In this primal state voluntary movements look extremely awkward, exactly like the inept behaviour of a small mammal that picks its way along a particular path for the first time.

If we wish to define voluntary movement in terms of function, we must add a further consideration — namely, that it must be capable of being activated at any time. This is not the case with all locomotor patterns. Holst has demonstrated that the endogenous stimulus-production on which a particular behaviour pattern depends is in constant correlated ratio to the average frequency with which the pattern is performed by the animal in its everyday life. Thus the wrasse swims practically the whole day, and in normal conditions its pectoral fins hardly stop moving from sunrise until it goes to sleep shortly before sundown. The sea horse, on the other hand, only swims on average a few minutes a day. If the wrasse's brain is removed and artificial respiration applied, its pectoral fins beat continuously; under the same conditions the sea horse's dorsal fin, its chief locomotive organ, does not move at all. However, it does not rest in its normal position in the groove provided for it on the animal's back, but stays in a half-erect position. By applying certain stimuli, such as pressure in the neck region, one can induce the fin to return to its normal position of rest, but when one relaxes this pressure after a while, the fin raises itself even higher than it was before, and the longer the stimulus has kept it in its folded position, the higher it rises. In the end it not only rises to its fullest extent but also performs for a time the undulating forward swimming motion. C. S. Sherrington called this phenomenon 'spinal contrast'. Holst interpreted it as follows: the erection of the fin is fed by the same endogenous stimulus-production that causes its swimming motion, and consumes the same kind of specific energy; the motor activity of the erection of the fin has a lower threshold than the swimming motion and consumes less of this energy. The semi-erect position of the fin in the unstimulated spinal cord preparation consumes exactly the amount of endogenous energy that is being continuously produced. The relaxation of the fin in response to the pressure in the neck region has the same effect as that normally produced by the higher centres of the central nervous system. As long as inhibition lasts action-specific energy is conserved, and when inhibition ceases the energy activates the motor pattern, which has a higher threshold. This interpretation is confirmed by the fact that after it has finished beating the fin in the spinal cord preparation gradually returns in an asymptotic curve to its previous semi-erect position.

One finds similar differences in the action-specific energy production of common and less common behaviour patterns among many animals. Small birds like finches and tits, for example, change from hopping to flying and vice versa innumerable times a day, and although they usually fly only short distances, they spend a considerable part of their waking life in flight and must always be ready to take off. To the observer their behaviour has the character of a voluntary movement, for the bird can never find itself in the position of 'wanting' to fly and not being able to do so.

But with birds that fly only rarely, like geese, this can happen. Except when migrating, geese usually fly only twice a day, in the morning and in the evening. Even if one conditions them to fly at different times of the day, one produces a situation in which they use this behaviour pattern — which is now no longer set in motion just 'for its own sake' — in a similar way to that in which we employ voluntary movements. Their actual take-off, however, now becomes very different from what someone counting on voluntary movements would expect. When the keeper first calls out as he moves towards their feeding place the birds immediately become interested and begin to walk slowly but surely to the spot from which they usually fly off. Yet once they reach this spot they do not at once spread their wings and take off, but stop, stretch their necks and begin an elaborate procedure of 'getting themselves into the mood' for flying. The sounds they make slowly change, becoming shorter and more staccato until they blend into the typical take-off call; at the same time, they begin, with increasing frequency, to perform their pre-flight movement, shaking their beaks from side to side. This signifies that the bird is now in the mood for flying and passes the mood on from bird to bird. Finally the goose frees its wings, crouches for the jump, spreads its wings and takes off — or else does not. For the whole process of increasing flying excitation can come to a halt at any point, whatever degree of action-specific arousal has been reached, and go into reverse. On a number of occasions I have seen a goose flex its legs and spread its wings, only to 'freeze' in this ridiculous posture for several seconds, looking as though it had been badly stuffed, before furling its wings and returning to an upright position.

A practised observer can tell from the speed with which the

goose's agitation grows whether it is going to take off or not. If it runs through the early, low-threshold phases of rising excitation quickly, one can extrapolate the rising curve and predict that the maximum degree of excitation will be reached. If the rise is slower and its curve shows the tendency to level out, one can again extrapolate and predict that it will not reach the threshold of actually flying. For reasons as yet unexplained the curves described by the rising or falling of action-specific excitation never show any sharp change of slope, as if the changes of arousal obeyed their own law of inertia.

The human observer often becomes impatient while watching processes of this kind. When one sees a goose struggling for minutes on end to rouse itself to the point of taking off, one has the same urge to help it reach the critical point as one does in the presence of someone trying to sneeze. A dog lover abhors the popular habit of teaching a dog to beg because it makes the animal perform a non-voluntary action with a threshold which it has to struggle painfully to attain.

Among locomotor coordinations the rarely used ones like that of the sea horse swimming, or of the greylag goose flying, are in fact the exception, in that most animals have an unlimited number of locomotor patterns at their disposal. A mallard drake cannot perform a courtship movement, or a rooster crow 'to order', any more than a man can sneeze when he is told to, but all three can walk one step forward whenever necessary. The necessity to be able to do this is obvious, and it is surely for this reason that the vast majority of voluntary movements have emerged from the stock of fixed motor patterns contained in the act of walking. When an animal 'wants to do something it cannot do', it almost invariably makes, or begins to make, walking movements or parts of such movements. When a dog looks up eagerly at the bowl of food that its master is bringing into the room, it lifts its front paws up and down one after the other; a horse in the same situation paws the ground with its hooves, and so on. It is understandable why units of locomotor coordinations should be the commonest components of learned behaviour patterns.

As we have already said, proprioceptor mechanisms certainly play an important part in the creation of motor skills, but certainly less once the pattern has been completed. A true voluntary movement — that is, a new putting together of voluntary elements — always appears remarkably awkward — something also referred to above (p.101). The processes of reafference which have to check and recheck the newly acquired pattern clearly take a considerable time.

This is a vital point. These reafference mechanisms in the learning of motor patterns are of decisive importance for the development of the central representation of space, the faculty which underlies all higher kinds of insight-controlled behaviour. The two processes of motor learning and of the acquisition of knowledge through feedback have evolved hand in hand. From the phylogenetic point of view it was probably the survival function of motor skills that caused the selection pressure from which voluntary movement proper came into existence. The possession of a behaviour pattern adapted to specific spatial parameters, which runs its course with the speed of a well-oiled machine, without any delays due to reaction times, can be a matter of life and death for many vertebrates. But it only needs a slight shift of emphasis, such as must have occurred with the curiosity behaviour of higher animals, and particularly the self-exploration of our immediate ancestors, to bring the survival value of the acquisition of knowledge to the fore.

The faculty which originally had only helped to produce the motor skills became an important means of exploring. A baby's exploratory play is at least as important for the formation and development of the child's central representation of its spatial environment as it is for the learning of motor skills. As the research of T. G. Bower and W. Ball has shown, our conception of space is not derived solely from the evidence provided by our sense of touch. An infant's size-constancy mechanism is in operation long before the infant begins to explore space with its hands. But it is reasonable to believe that the interplay of motor learning and the development of a conception of space is of vital importance for learning the shapes of physical objects. If our voluntary movements, composed of numerous combinations of minute motor components, were not able to reproduce all spatial phenomena (size permitting), our sense of touch could not become the important source of experience of space that it is.

The intimate connection between these two functions is also seen in the fact that the human organ which is capable of the most subtle voluntary motor activity, namely the index finger, has in relation to its size the largest representation in the sensory area of our cerebral cortex. The tongue and the lips also occupy a strikingly large area

— larger than that for the whole hand. A similar situation is found in chimpanzees. Mouth and tongue were obviously the most important tactile organs among our mammalian ancestors before the anthropoids transferred this role to the hand. Much the same thing happens in an infant that puts everything new into its mouth in order to find out what it is. We also know from our adult experience that our tongue can give us remarkably exact impressions of size and shape, often slightly enlarging real parameters: the cavity in a tooth seems enormous.

This function of voluntary movement to acquire information about external objects by feedback is a particular instance of a general principle. All acts of exploration basically consist of actions performed in order to collect their feedback. A motor skill that is formed to fit a particular spatial structure performs the same functions, but in its own way, as follows: in the course of the learning process through which, as we have seen, it was assembled from individual components, it supplies from a correspondingly large number of details an image of the object; each time the motor skill is performed the image is compared with the reality and any discrepancies fed back and corrected. This is a typical example of what is known as pattern matching (see above, p.24).

A similar process takes place on a higher plane when a pattern is controlled by several sense organs. I have already referred to the significance of the fact that the anthropoid's grasping extremity remains within its field of vision, with the result that the exteroceptor reafferences of vision coincide with the proprioceptor perceptions of the position and movement of the limbs, thus producing the cognitive act of pattern matching. When a baby discovers its own hands and feet and starts to play with them, not only is the number of reafferences doubled but it becomes abundantly clear that they come both from inside and outside the subject.

7.6 *Curiosity and self-exploration*

In a broad sense, and in purely functional terms, all behaviour can be called exploratory in which an organism does something in order to learn something. Such a definition would include all motor activity that feeds back adaptive information through sensory channels. This occurs in the cases where, as described on p.92, an animal tests out one and the same behaviour pattern in different situations or on different objects, like the jackdaw discovering suitable material for its nest. Again it must be emphasized that the motivation for this trial-and-error behaviour arises entirely from the urge to perform that particular motor pattern.

It is probably from this kind of learning — 'operant conditioning', as the behaviourist school calls it — that a far more effective form of exploratory behaviour has phylogenetically developed. It differs from the preceding form in two vital respects. Firstly, it is not *one* fixed motor pattern that is tried in various situations and on various objects but virtually all the patterns in the repertoire of the species are exercised successively on one and the same object. Secondly, the behaviour is not motivated by the urge to perform a single purposive consummatory act but derives from a different source, which has the remarkable power of being able to activate many, if not all, of the animal's innate fixed motor patterns. Monika Meyer-Holzapfel was the first to understand clearly the peculiar characteristics of this process, which we call exploratory or curiosity behaviour.

It is clear that when birds and higher mammals are at play, their rapid succession of instinctive behaviour patterns cannot derive from the same sources as motivate these patterns when the situation is a serious one. When a kitten is playing, for example, it performs actions taken from the contexts of catching its prey, challenging its rivals and defending itself against larger predators, all in the space of a few seconds, without pause and in completely planless succession. But if a cat is menaced by a dog and arches its back in the familiar posture of defiance, it takes many minutes, sometimes half-an-hour or more, to calm down and attain a state in which it can consider other activities, such as chasing its quarry or fighting its rivals. Monika Meyer-Holzapfel's conclusion that there must be two different sources of motivation here seems to me convincing.

It is difficult to decide whether what we call a young animal's 'play' derives from exploratory behaviour in the strict sense, or whether it is the other way round. There are all manner of intermediate stages between the two. The nature of exploratory behaviour becomes clearer, the greater the number of different behaviour patterns that are tried out on the same object or in the same situation. If one offers a young common raven a completely unfamiliar object of an appropriate size, it first reacts in the way that an adult bird 'mobs' a potential predator: it approaches cautiously, hopping sideways, then finally delivers a sharp blow with its beak and immediately flies off. If one paints two dummy eyes on one end of an oblong object the bird will attack the opposite end. If then the object does not chase it, as a real predator would, the bird takes up the attack itself, as though it were dealing with a quarry that was able to put up a stout defence, striking with its beak at the eyes or the 'head'. Once the object is 'dead', the bird uses all its available instinctive behaviour patterns to pull it to pieces, examining it to see whether it is edible. Finally it hides the pieces. Later, when it has become indifferent to the object, the bird will occasionally hide beneath the pieces other objects which have since become more interesting to it, or use some of the bigger pieces to perch on.

These are precisely the consecutive events which Arnold Gehlen described as exploratory behaviour: an object is first examined and 'made familiar', then discarded in the sense of being 'filed away' in a manner that makes it possible for the animal to return to it if necessary.

Until this need arises, one cannot observe anything in the animal's exploratory behaviour to suggest that it has learnt anything. When a brown rat goes through a process of exploration similar to that of the raven, running and clambering over all the possible paths in its area, it does this until it knows every single path that leads to a hideout, and what protection each hideout gives. But this mass of information only comes into play if a strong key stimulus causes the rat to take flight; up to that point, what it has learnt remains concealed. Hence the concept 'latent learning' — although the learning process itself, as we have seen, is fully apparent, and it is only the knowledge acquired through it which is latent, and this only as long as it is not needed.

As with the behaviour patterns of play, so also those of exploratory behaviour are not caused by the specific motivations which otherwise activate them. It is not only that, as in play, the patterns follow one another far more quickly than they ever can in 'serious situations'. There is another reason to think that all exploratory behaviour has another common motivation: that is, that it ceases immediately when any mood other than that of inquisitiveness is activated. The raven in our example will show behaviour patterns of escape, of chasing its prey, of eating, and so on, but will interrupt them at once if a real situation of escaping, chasing or eating arises. When it really gets hungry it immediately ceases to explore and goes to its food dish or begs from its keeper;

in other words, when things become serious it will fall back on objects and behaviour patterns that it already knows will appease its hunger. Gustav Bally was the first to show specifically that play was only possible in what Kurt Lewin termed a 'tension-less field'.

Gehlen, in his motivational 'field theory', accurately character- ized the nature of exploratory behaviour when he described it as consisting of 'senso-motory patterns of combined visual and tactile sensations, cyclic processes that provide their own stimuli for their continuation. They run their course with no hint of desire and serve no immediate consummatory purpose. This form of productive interaction with the outside world is, at the same time, objective.' One could hardly find a better definition of exploratory behaviour. It clearly distinguishes learning by exploration from the ordinary mechanisms of the operant acquisition of conditioned responses described in the earlier sections of this chapter. Of these latter Gehlen says: 'It is only the pressure, in a given situation, of a present instinctive urge that forces learning processes to operate, making the animal's behaviour essentially dependent.... And since its actions are not independent, they are not objective.'

In the book from which these quotations are taken Gehlen made the mistake — which he later corrected — of attributing objective exploratory behaviour solely to human beings. I therefore repeat: exploratory behaviour in the true sense is absolutely objective. The raven that investigates an object has no wish to eat it; the rat that examines all the nooks and crannies of its territory has no wish to hide; they both want to know whether the object in question *can* be eaten or used as a hiding place. Uexküll once said that all objects in an animal's world are 'objects of action', i.e. things with which something has to be done. All objects that have been explored and then 'filed away' in this manner have been objectivated in a higher sense, since the knowledge of how to employ them has been both acquired and remembered independently of the pressure of the ever changing motivational situations within the organism as well as of the environmental situations around it.

Animals that are capable of learning the characteristics constantly inherent in the various objects of their environment are, understandably enough, adaptable in a very high degree. By treating every unfamiliar object as though it were biologically relevant, they do in fact discover all the objects that are. This enables the common raven, for instance, to live in various biotopes as though it were specially adapted to each of them: in the North

African desert it lives as a carrion vulture, on the bird islands of the North Sea it lives as a skua, parasitically feeding on the eggs and the young of the breeding colonies, and in central Europe it manages to survive by preying on small creatures in the manner of a crow.

The phylogenetic programmes of exploring animals are always extremely 'open', to borrow Ernst Mayr's term. The knowledge of the objects or processes that make up their environment is not specified in innate release mechanisms containing a wealth of information and characteristics, but is acquired by objective investigation. Typical exploratory animals are also 'open to the world' — a quality which Gehlen regards as one of the characteristics that distinguish man from the animals — albeit not to the same extent as man; nor, in animals, is this quality integrated with the other prerequisites of objective thought discussed in this chapter to form a higher system.

A behaviour programme that can be modified within such broad limits as those of exploratory animals requires motor patterns that can be used in a variety of ways. The same is true of the organs involved. Higher morphological specialization of organs excludes diversity in their utilization. Hence from the morphological point of view all characteristic exploratory animals are comparatively little specialized representatives of their taxonomic groups — 'specialists in non-specialization', as I like to call them: rats from among the rodents, corvids from among the song birds, and man from among the primates. It is also significant that among the higher animals only such 'non-specialists' can become cosmopolitans. Certainly a rat or a human being may, physically speaking, perform less efficiently on occasion than an animal highly specialized in that particular field, but both will outstrip their closest zoological relatives in the versatility of their motor skills. If human beings were to challenge the entire animal kingdom to a test of versatility, consisting, say, of walking twenty miles, swimming fifteen metres underwater at a depth of five metres and simultaneously retrieving various specified objects, then climbing several metres up a rope — all of which any average person can do — there would not be a single other mammal capable of doing all these three things. If one were to substitute a tree for the rope, the polar bear could match a human being; so could certain macaques if the distance to be walked or to be swum under water were reduced. Provided that one scaled down the requirements to suit its size the only real rival in all three tests would be the rat.

When one considers the consequences of exploratory behaviour, particularly its importance for the evolution of abstract thought, one is surprised by how trifling the changes and subsequent integrations seem to be when regarded severally and without reference to their revolutionary results. None of the identifiable mechanisms involved as subsystems in the acquisition of conditioned responses needs to be substantially changed in its function and none becomes superfluous. The new 'discovery' simply consists in generalizing appetitive behaviour in such a way that its goal is not the elicitation of a specific instinctual consummatory act but the learning situation itself. This requires only a shift of emphasis, for almost all appetitive behaviour in higher animals is accompanied by learning, by the acquisition of conditioned responses. This already emerged from our discussion (pp. 21-5) of processes in which the information is encoded in the form of the fixed motor pattern, and the nature of the appropriate object is discovered by trial and error. What is new in exploratory behaviour is only that the motivation is furnished by the learning process itself, not by the achievement of the final consummatory state.

As a result of this apparently small step forward there emerges an entirely new cognitive process which is in essence identical with that of human investigation and which leads without an essential break to the activity of scientific research. The connection between play and investigation remains consistently close in the course of this development, and is still fully preserved in adult human investigation, whereas in adult animals it tends to disappear. 'Man is only completely human when he plays', said Schiller. 'In every true man there is a child hidden', said Nietzsche — to which my wife once added: 'Hidden? What do you mean?'

As I have shown in detail elsewhere, the retardation of human development, the partial retention in later life of an adolescent stage of development — the process known as neoteny — ensures that man, unlike most animals, does not break off his exploratory behaviour when he reaches adulthood but retains his constitutional openness to experience into old age.

As we already know, gestalt perception, though evolved as a 'constancy mechanism' in the service and under the selection pressure of recognizing individual objects (see p.116ff.) is proved able to arrive at 'abstracting' the objective laws underlying a variety of individual phenomena. Similarly, exploratory behaviour, without any substantial mechanism, also performs a new function which in embryonic form is found in the animals closest to man, but becomes an essential and constituent faculty only in humans — namely, self-exploration.

One may well ponder how and when our ancestors became aware of their own existence. An animal for which exploratory behaviour was one of the activities vital to survival could hardly fail to discover its own body sooner or later as an object worth investigating. The fact that it was anthropoids that made this step forward is the result of circumstances that we have already encountered, i.e. that, as climbing animals with grasping extremities, they have a well-developed faculty for the central representation of space as well as for highly sophisticated voluntary movements. Moreover, their grasping extremities are continuously within their field of vision, which is not the case with the majority of mammals, including many apes. When a dog puts his front paw down on a spot which he had seen a fraction of a second earlier, he cannot see his foreleg or his own body moving; virtually the same is true of vervet monkeys, macaques and baboons. The deliberately climbing anthropoid ape, by contrast, sees its hand almost the whole time, together with the object in its grasp, and this is especially true of non-locomotive but exploratory movement. In humans, as Mittelstaedt has shown, the direction in which one stretches one's arm and hand is perpetually checked and corrected by the feedback supplied by the eye with far greater accuracy than the proprioceptors of our depth-sensitivity can achieve.

As far as I know, there have been no comparable experiments with anthropoid apes, but I once observed a chimpanzee lying on its back and cautiously covering first one eye, then the other with its hand as the light from an electric bulb shone on it. It looked as though the chimpanzee were exploring the results of its own behaviour, but be this as it may, its hand and the object in question were both in its field of vision. This faculty gives the anthropoids the opportunity to observe the interaction of the two, as does also the act of social play. This latter activity takes up a great part of a young ape's waking life and becomes, like all higher exploratory behaviour, a kind of dialogue, in which a question is asked of an object and the answer recorded. When two curious young chimpanzees play with each other this interaction doubles; if one of them takes the other's hand and studies it intently, as young chimpanzees quite often do, all the premises are there for the pioneering discovery that its own hand is essentially the same as his brother's. It seems to me very likely that the anthropoid realizes the objective nature of its own body in the mirror image of its species-specific playmate.

The animal's discovery of its own body, particularly of its own hand, as something to be explored, need not necessarily be regarded as true reflection. As yet, it does not arouse the sense of awed surprise that is usually considered as the starting point of reflective thought and of philosophy. But the sheer realization of the fact that its own body or its own hand is an objective 'thing', with fixed distinguishing characteristics like any other such 'thing' in its environment, must have been a momentous happening of truly revolutionary significance. Insight into the objective nature of one's own body and its effective organs leads of necessity to a new and deeper understanding of the interaction between one's own organism and the objects in one's environment. Once understood as a separate entity, one's body becomes a measure for comparison and thus makes it possible to assess the outside world in terms of the self.

This revelation opens up to the organism a true shock of recognition, the objectivation of its environment. For the moment one of our ancestors recognized for the first time that its own hand and the object it held in its grasp were both objects in the material world, and realized the relationship between them, its knowledge of the grasping process became conceptual thinking, and its awareness of the object that it grasped became a concept. Incidentally, the Latin word *concipere* means 'to grasp'.

This process is, of course, closely connected with, and dependent on, the other processes described in this chapter. Thus the function of abstraction involved in perception enables the animal to recognize an object which it explores under the conditions of perception. Likewise the central representation of space, with all the subservient functions that go to make it, and especially the inexhaustible fund of knowledge derived from the reafferences of voluntary behaviour, are indispensable conditions for self-exploration.

7.7 *Imitation*

Imitative behaviour is an activity that can only be called cognitive in a very broad sense. However, it is a prerequisite of abstract thought in that it is essential to the integration of the functions described in the course of the present chapter with the process of tradition discussed below. From the phylogenetic point of view it probably arose from the play and exploratory behaviour of social animals with well-established family cohesion. Its own prerequisite lies in the interlinked functions of voluntary behaviour and its proprioceptor and exteroceptor control.

Imitation is, of course, one of the preconditions for learning verbal speech and, thus, indirectly of numerous other specifically human achievements. It is a matter for debate why it should be the mouth that became the organ of this communication system. We have already seen that the lips and the tongue became capable of particularly subtle voluntary behaviour among our ancestors and provided a wealth of informative reafferences. Their motor functions, together with those of the larynx, also considerably affect the production of emotionally expressive sounds and gestures, and in all social intercourse among animals both closely and distantly related to man, the face, and especially the mouth, is the part of the body on which attention is chiefly fixed, and it was bound to become the principal instrument for the emission of signals.

Yet the actual process of imitation remains a mystery as concerns both its physiological origin and its distribution over the animal kingdom. Strictly speaking, the ability to imitate is only found, apart from man, in certain birds, especially song birds and parrots — though here confined to vocal imitation. The imitative powers of apes are a familiar talking point, but even anthropoids only show an incipient ability to give a close imitation of a particular pattern of behaviour, and their imitative powers are nowhere near as accurate as those of birds. A chimpanzee will understand at once what is happening when it sees a man opening a door with a key, and after a few minutes it succeeds in doing the same. But the 'aping' of a movement or a facial expression just for the sake of imitation is something of which, as far as I know, there are only faint traces in apes.

Children, on the other hand, and, astonishingly, the above-mentioned birds, are excellent imitators. Social psychologists have proved that children imitate the actions of adults with great precision out of sheer pleasure in imitating, long before they understand the purpose of the behaviour pattern in question. In their book *The Social Construction of Reality*, Peter Berger and

Thomas Luckmann have analysed these processes very closely. Children display a particularly well-developed awareness of ritualized movements such as form part of social exchanges. When he was barely two years old my eldest grandchild was once greatly impressed by the way a Japanese friend of mine bowed, and he imitated this bow in what could fairly be called an 'inimitable' manner. No adult could have imitated it so perfectly at the first attempt. It was a highly comical episode, but it was only because my friend was an ethologist that he did not feel that the child was 'aping' him.

The exactness with which men imitate sounds and movements is of very great social importance, because it is minute similarities and dissimilarities of accent and of manners that sustain the cohesion as well as the separation of social and ethnic groups.

Some song birds and parrots, especially in certain phases of their ontogeny, show a marked predilection for complicated patterns of sounds which are only just within the scope of their faculty of imitation. When its keeper whistles to a young bullfinch, it quickly comes to the front of the cage, cocks its head with one ear towards the source of the whistling, ruffles its ear feathers and listens, fascinated. Starlings behave in the same way, and young Shama thrushes (*Copsychus malabaricus*), listening intently to their father's lovely song, are a charming sight.

In many cases it is months before what the young birds have heard finds its way into their own motor vocalization. Heinroth tells of a nightingale which heard the song of a blackcap during the period between the twelfth and nineteenth day after it had hatched. When it began to sing, about the following Christmas, it reproduced the blackcap's song with the same accuracy as the recording Heinroth had made at the time. The bird must have retained, purely acoustically, a memory of the song that it was able to transpose into movement months later.

The researches of Konishi have shown how this transposition from the sensorium to muscle movement takes place. In certain species the song is innate in so far as even a young bird confined in sound-proof surroundings develops it almost normally. However, it is not a fixed motor pattern of the song that is inherited: the bird possesses an innate acoustical pattern or 'template' of the song. By trial and error, producing many kinds of sound utterances, the young bird gradually develops a sound pattern matching this template. Heinroth suspected this to be the case when he coined the term 'self-imitation', and Konishi proved the case by removing, in species of this type, the auditory organs of very young birds. The birds could only produce an amorphous twittering, without any pure notes, and the profusion of strong overtones gave the impression of mere noise. Unlike the song, call notes and warning calls are based on innate fixed motor patterns in all song birds hitherto investigated whose song develops in the way just described. All these signals are emitted in a completely normal way even by birds deprived of their auditory organs. The same is the case with all sound utterances of chickens, ducks and most other birds incapable of imitation.

The acoustic 'template' mentioned above has two different, though often overlapping functions in most cases. Only rarely is it so perfect that it can provide the young bird with complete information about how its species-specific song should sound. A bird which is brought up in sound-proof conditions but has not been deprived of its sense of hearing, generally produces a simplified but still recognizable version of the song proper to its species. It can be assumed that in natural conditions the innate 'template' would tell the young bird which of the many bird songs around was the one it had to imitate.

Many years ago I had a feeling that this must be the case when I observed the behaviour of a house sparrow. Heinroth and others had found to their surprise that in spite of its simplicity the chirping of the sparrow was not innate but had to be learnt by imitation, like many more complicated bird songs. F. Braun has shown that sparrows brought up together with goldfinches learned the complex song of the goldfinches without difficulty. When I reared a male sparrow that had been taken from the nest when it was two days old I wondered which of the many bird songs to be heard in my room it would imitate, but to my disappointment it just chirped, thereby apparently contradicting what a number of famous ornithologists had said. After a while, however, I realized that it was not chirping like a sparrow but like a budgerigar. Surrounded by many far louder and more striking songs, it had chosen to imitate the chirping of the budgerigar because this was the closest to its own acoustical template.

Whereas we know from Konishi's investigations that afferent — or more accurately, reafferent — control is a built-in component of the imitative process in the birds concerned, we know nothing at all about the physiological mechanisms of human imitation. It is my own firm conviction that in cases where physiology leaves us in the lurch we can and must call on the help of phenomenology, on self-observation as a legitimate source of knowledge. Strictly speaking, one can only do this in terms of oneself, its objective validity depending on the extent to which others are capable of appreciating and understanding the experience one is describing. Having said this, I will attempt to describe what I experience when I try to imitate something by muscle movement.

To begin with, the sight of some gesture or facial expression releases in me a 'primary' urge — i.e. one free of any other recognizable motivation — to imitate what I see, and this, surprisingly, I usually manage to do at the first attempt. The first time I tried to give an imitation of the East German leader Walter Ulbricht's hard-bitten expression, people greeted my success with peals of laughter, and there was no need to tell them whom I was impersonating (I should perhaps add that my beard probably helped to bring it off).

Strangely enough, my first experience in such cases is of a kinaesthetic nature: I seem to feel 'what it must be like' to make a particular grimace or perform a particular action. After I have repeated the action a number of times, the imitation improves to some small extent. To observe myself in the mirror does not help a great deal, for the proprioceptor reafference is sufficient to tell me when my imitation has reached the best possible approximation to my kinaesthetic image. Further repetition leads to no improvement but rather tends to destroy the original vision, which was the product of unconscious processes — or, more accurately, it tends to obscure this vision through the superimposition of successive recollections of my imitations, so that the original gets lost underneath the copies.

The very fact that small children are better at such impersonations than adults supports the idea that it has nothing to do with rational processes. So, too, does the fact that one cannot always perform or repeat such impersonations at will but has somehow to be 'inspired', as with the functions of higher Gestalt perception. The whole thing has something in common with the act of play, in that it can only take place in an atmosphere of cheerful relaxation. If one compares the imitative activity of small children with the way young birds learn to sing, one is struck time and again by the parallels between two processes which occur in very different creatures and probably have very different physiological roots.

The phenomenology of human imitation suggests that the first step is the emergence of a sensory model or prototype. This would probably strike us as highly unlikely if the above observations and experiments with birds had not shown that this was obviously the way their vocal imitation came about. At the same time the process, both in human beings and in birds, must be based on a highly complex sensory and nervous apparatus, such as we are accustomed to find only where its function has a marked survival value. As far as human beings are concerned, the question of the survival value of imitation is almost trite, and was dealt with at the beginning of this section. But we know absolutely nothing about the physiological processes in man that convert the sensory pattern into a motor pattern. It certainly does not come about by trial and error, as we can reasonably assume is the case with birds. And as far as these, our only colleagues in the art of music and imitation, are concerned, we know virtually nothing about the survival value of these powers of imitation, which are not present in any other sentient being. Mating calls, warning calls and similar means of communication, as mentioned above, are based on innate motor coordination.

7.8 *Tradition*

Among highly developed social animals we know quite a few cases in which knowledge acquired by an individual can be preserved beyond its life span by being transmitted to the community at large and handed down from one generation to another. We are so accustomed to think of a biological process whenever the words 'heredity' and 'inheritance' are mentioned that we are apt to forget their much older juridic connotation. The process by which individually acquired knowledge is passed on from individual to individual through long periods of time is called 'tradition'.

Such transmission of knowledge can take two forms: first, if an experienced individual reacts to a certain situation by uttering a warning cry, or induces its fellows to take flight by doing so itself, the escape response may become associated with the whole stimulus situation (see p.77ff), even after one single occurrence. Secondly, superior learning processes may cause an animal to re-enact the behaviour of its older fellows. This need not be imitation proper: it is sufficient for the exploratory behaviour of the inexperienced animal to be led in a particular direction by the example of its experienced fellows. I was the first, in the 1920s, to prove the existence of tradition in animals by my observations on jackdaws. My initial, unpleasing discovery was that hand-reared young jackdaws were not only very tame towards human beings, but were also not in the slightest afraid of dogs, cats and other predators, and were thus exposed to great danger when flying free.

The jackdaw's sole species-specific instinctive behaviour pattern, which protects it from predators, consists of an innate releasing mechanism and a single instinctive motor pattern which this mechanism releases. The mechanism responds to a stimulus combination which must present the following characteristics: something black and flexible must be carried by a living creature of some kind; a stiff and solid black object like a camera is no use, nor is something flexible and dangling that is not black. The soft black paper attached to film packs or a pair of wet, black swimming trunks carried in my hand released the response just as intensely as a struggling live jackdaw I had caught. The nature of the creature that plays the part of the predator carrying its prey in this scene is quite immaterial; a jackdaw flying to its nest with a black pinion from another corvid in its beak would elicit a full response.

The appearance of this stimulus combination causes every adult jackdaw in sight to emit a penetrating rasping, rattling sound. At the same time it bends forward in a strange way, and its outspread wings begin to tremble. This behaviour is apparent even on the wing. Every jackdaw within earshot flies rapidly towards the calling bird and joins in the rattling chorus, and when a large enough number of birds have assembled, they ferociously attack the 'enemy', not only lunging at it but digging their claws into it and raining down blows on it with their beaks. In the experiment the 'enemy' was my own hand holding a dummy jackdaw, and the back of my hand was soon covered in blood.

Even after the very first occasion when I unwittingly led the jackdaws to attack me, I was struck by the fact that they distrusted me for a long while afterwards, and I realized that, if I performed any more such experiments, there was a danger that the birds would lose the tameness that I required them to have. They were also used to seeing other jackdaws sitting on my arms or shoulder, which did not elicit the response described above. However, this was the result of a process of habituation, as emerged from the fact that some years later, when they had become too timid to land on my body, they immediately started their rattling noise when they saw me with a young tame jackdaw perched on my shoulder or the back of my hand. This reaction frustrated my plan to try and increase the average tameness of the jackdaws that had grown timid, by putting tame hand-reared birds among them, as can easily be done with wild geese.

Once a person or an animal has been set upon by a jackdaw while carrying a soft, flexible black object, he becomes, by a conditioning response, the stimulus that releases the rattling attack, which he still does even if he appears without the sinister black object. Crows behave in almost the same way. My friend Gustav Kramer used to take his tame carrion crow with him when he went for walks in the woods round Heidelberg. As soon as the bird perched on his shoulder the wild crows began to make hostile noises. In time Kramer became so hated that the birds mobbed him even when he came out of his house dressed in an office suit, which he never wore when he had his tame crow with him. The same thing used to happen to me whenever I tried to rear a tame jackdaw; the result was always the reverse of what I had intended.

Since young social corvids follow their parents faithfully, and normally never leave the immediate vicinity of their nest site, where in the case of jackdaws there is always a crowd of older birds, it scarcely ever happens that a young bird comes across a dangerous predator without an experienced bird being at hand and reacting with its loud rattling response. As Sverre Sjölander could show, thanks to his gift for imitating bird calls, it only needs a cat or other animal to be shown to a young jackdaw once or twice, and the rattling sound to be made at the same time, for the bird to have the fear of the cat implanted in its mind for the rest of its life. Sjölander even used this method to develop in his jackdaw an antipathy towards fellow jackdaws and other corvids, and to prevent it from flying off with the migratory birds in the autumn.

In geese, tradition has another important role to play, viz. in migratory navigation. Geese reared artificially, since they are resident birds, usually stay in the place where they have been brought up. My experience with the colony of wild geese I had transplanted to Seewiesen showed that, without being guided, such young birds would not dare to land in an unfamiliar place, and to get them to settle in fields that we had leased for this purpose outside the fence of our field station, it became necessary for the human foster mother on whom the geese had been imprinted to go into these fields with them and to feed them there patiently in the hope of advertising the place as a grazing area.

Once acquired, path habits remain firmly fixed in geese, as the following example shows. In the 1930s I had four greylag geese in Altenberg which used to accompany me when I went out into the broad flood-plain of the Danube, where there were very few trees. I used to cycle along the dam that encircled the area, while the geese circled above my head or occasionally landed nearby. To reinforce the stimulation further, I used to make for certain reedy overgrown pools where I knew the birds would feel at home. They enjoyed these excursions as much as dogs enjoy their daily walk, and they used to wait in front of my house at the accustomed hour, generally flying off the moment I took my bicycle out of the shed.

I then wanted to see what the geese would do if I did not leave at the usual time. From the roof of the house I could see with my binoculars the area where we usually went. As the accustomed departure time came, the geese grew restless and made increasingly louder pre-flight noises, lasting much longer than if I had left with them on my bicycle as usual. Finally they took off, making first for a spot in a nearby field where we used to meet after I had cycled through the village — for the birds were afraid of the street and avoided it. Having circled a few times above this spot, calling loudly, they made off towards the pool where we had spent the previous afternoon. They circled above the pool for some time, calling as they did so, then flew off to a second pool, which we did not visit so frequently. When they did not find me there either, they flew even further away and looked for me by a pool in a gravel pit where we had only been very rarely, and not for a long while. Here too they circled a few times, then returned to my garden without having landed anywhere the whole time.

These observations make it entirely plausible that not only the general course of the migration route is passed down from generation to generation, but also the knowledge of each individual landing place. This is supported by everything we know from other sources about these birds' retentive memory, as well as by the fact that, as Dutch field ornithologists discovered, there are certain lakes and ponds where, during migration, flocks of greylag geese of more or less equal number assemble, which they assume to be the same geese, together with their young; they arrive and depart each year at approximately the same dates.

Steiniger discovered a tradition in brown rats that extended over several generations. The knowledge in question concerned certain poisons. Experienced rats indicated the presence of a poison by urinating on the bait, though even just avoiding it also seemed to act as a warning. Rats and other omnivorous animals usually begin by ingesting only very small quantities of unfamiliar foods (cf. above, p. 91), and this cautious method may be one source of the knowledge which is then passed on.

Observations carried out on macaques by Kawamura, Kawai, Itani and other Japanese scientists have shown that there are traditions of motor behaviour which are invented by one particular animal and then, having proved to be 'profitable', have spread to the entire colony. The process could be followed in detail. In one such case it turned out that a particular animal, a young female, made a number of discoveries. First she introduced the habit of washing dirty yams in a stream. After a number of macaques had followed suit, a few tried washing the yams in sea water; finding that this gave them a pleasant spicy flavour, they also dunked them in the water while they were eating. When the animals were fed with wheat which was simply scattered over the sand of the sea shore where they lived, the same female that had started washing the yams now began to throw the wheat, and the sand underneath it, into the sea. Initially this was probably an automatic extension of the procedure that had proved profitable with the yams. But, as so often, the application of a false assumption unexpectedly produced a successful result for the sand sank to the bottom, leaving the grains of wheat floating on the water. The procedure was basically the same as that of panning for gold. At the time these observations were carried out nineteen of the macaques had adopted it.

There is one vital respect in which these examples of animal tradition differ from human tradition: they are all dependent on the presence of the object with which the tradition is associated. An experienced jackdaw can only tell an inexperienced jackdaw that cats are dangerous when a cat is actually there to demonstrate the fact, and a rat can only teach its inexperienced fellows that a particular bait is poisonous when the bait is actually present. This seems to be true of all animal tradition, from the simplest transmission of conditioned responses to the most complex learning by imitation. This dependence on the presence of objects is probably the obstacle which prevents animal tradition from *accumulating* in the way it does in Man. A specific tradition, such as that of the jackdaws' knowledge of cats, is broken once the object on which it depends fails to appear in the course of one particular generation, and the fact that all animal traditions are thus comparatively short-lived may well prevent their joining up with each other and creating a fund of common knowledge. It is only the development of abstract thought, together with the complementary development of verbal language, that enables tradition to become free of objects; for by means of independent symbols, facts and relationships can be established without the concrete presence of the objects themselves.

7.9 *Summary*

In Part Two of Goethe's *Faust* Helena says:

> *Doch red ich in die Lüfte,*
> *Denn das Wort bemüht sich nur umsonst,*
> *Gestalten schöpf'risch aufzubauen.*

> (But I am talking to the wind,
> for language strives in vain
> to conjure up shapes.)

The first part of this chapter is concerned with the methodological difficulties arising from the fact mentioned in the above quotation. In attempting to put complex systems into words, and to convey to the reader at least something of what I feel I know about them, I am perpetually haunted by the fear of 'talking to the wind'. Nowhere has this fear been stronger than in the present chapter, in which I have described a series of processes that are all prerequisites of conceptual thought, and thus of the rise of man. My task was made the more difficult by the fact that these various processes are not of equal significance and cannot just be linked together to produce a superior pattern or system, as one can plait osiers to make wicker baskets, but are of varying degrees of importance, sometimes closely, sometimes loosely connected, sometimes highly interdependent, sometimes less so. In addition, these relationships needed to be discussed in this chapter because they also have a part to play on the pre-human plane. I shall return to them, moreover, in the next chapter, where I shall discuss how all these functions are integrated to form a system of a higher order. The principles of this integration were sketched in the first section of this present chapter. Repetitions will be unavoidable.

The second section dealt with the function of Gestalt or form perception. All constancy functions, like those of colour, size, distance and shape, are mechanisms of *abstraction*, because they remove the purely contingent properties of the stimulus data from the process of acquiring knowledge. At the same time they are mechanisms of *objectivation*, because they convey consistent information, independent of the vagaries of the temporary conditions of perception, about the unchanging characteristics of objects. The ability to separate essential from inessential derives from sensory and nervous processes which are beyond our observation and rational control, but which in functional terms are very similar to the activities of making rational calculations and drawing conclusions. These unconscious processes are what Egon Brunswik called 'ratiomorphous' computing mechanisms.

Mechanisms of this kind are already to be found in relatively primitive animals, and from the phylogenetic point of view they have always arisen in order to help identify specific objects in varying sets of circumstances. In their most developed form, however, these 'computers' can bring into relief fundamental characteristics that constitute the common basis of a large number of specific objects.

The function of abstraction that Gestalt perception performs is undoubtedly not only a prerequisite, ontogenetically and phylo-genetically, for the development of abstract thought, but also remains an indispensable part of it.

Insight-controlled behaviour is directed to the survival-purposive solutions of problems by means of the mechanisms that convey instantaneous information (see Chapter 1.5). The most essential of these mechanisms are those of spatial orientation, of which, among the higher vertebrates, the most important are those of sight. The great majority of vertebrates use the parallactic displacement of retinal images for their general orientation and in order to avoid obstructions. Binocular focusing serves primarily to locate the animal's prey; only when the organism requires to locate inanimate objects in its environment very precisely does it focus on these objects with both eyes. The more varied the spatial characteristics of an animal's environment, the more accurately it needs to orientate itself; thus animals living on the ground or in trees need to orientate themselves more accurately than those that live in the water or can fly, and climbing tree-dwellers need this faculty most of all, particularly those relying on hands which grasp the branches or rocks.

In fishes which focus on objects, monocularly or binocularly, the process of visual orientation precedes locomotion. On a higher plane, a similar principle is realized in mammals, which first survey the situation for some time, in order to apprise themselves of the structural details of their surroundings, and then proceed to solve the problem posed by it at one stroke. Probably a kind of inner action takes place in a 'visualized space' — that is, in a model representation of the present environment within the central nervous system. All human thought can be seen as 'action' in a 'visualized' space of this kind, taking place within the central nervous system, a view supported by the linguistic evidence recently adduced by Porzig, Chomsky, Höpp and others.

Then the relationship between insight and learning was examined. Memory must have a part to play wherever a creature collects information about its environment before it performs an action. In particular, the more complex forms of insight-controlled behaviour require a large basis of information previously acquired by learning.

Contrariwise, there is scarcely any form of learning that is not, from its very beginning, guided by orientation processes. Even an inexperienced animal never acts in a completely haphazard unorientated manner when faced with a problematical situation, but is always directed by the instantaneous information it receives.

Finally, learned behaviour patterns that have firmly established themselves may thwart the solution of new problems. The history of science can show us examples of this.

The subject of voluntary movement was dealt with in section 7.5. It is the motor correlate to the sensory mechanisms that provide detailed spatial information during exploratory activity. There are only limited possibilities for this knowledge to exercise an influence on the motor side, and if they are to serve the interests of survival, the cognitive functions of orientation must be matched by a corresponding motor capacity. Phylogenetically both sides have developed hand in hand, and both perception of space and adaptability of motor activity are closely related to the demand made by the structure of the environment in which the animal lives. In relatively homogeneous environments, such as the open sea or the prairie, these demands are modest, while in mountainous terrain, on coral reefs or high in the trees, they are at their greatest. The motor pattern adapts itself to these requirements by making smaller and smaller units of locomotor activity available, the smallest being found in so-called voluntary movement. These units are combined by learning processes to form well-adapted sequences (motor skills, in the sense used by H. Harlow), the basic survival value of which lies in the speed, unretarded by reaction time, with which they are performed.

In the service of exploratory behaviour voluntary movement develops a new important function consisting in the feedback information on the spatial parameters by way of reafference. Consequently, those organs which are most precisely controlled by voluntary movement, such as the index fingers, the lips and tongue, have the largest areas of representation, relatively speaking, in the sensory part of the cerebral cortex. As a tool of imitation, voluntary movement is a prerequisite of verbal speech and, therewith, for the higher evolution of abstract thought.

Section 7.6 dealt with curiosity or exploratory behaviour. Certain animals are irresistibly attracted by unfamiliar objects and apply virtually all their behaviour patterns to them in quick succession. These rapid changes show that the individual patterns are not activated by the same motivations as are at work when the single patterns are performing their 'serious' function for survival. The very moment that the specific motivation (hunger, fear, etc.) arises, exploratory behaviour ceases. It can only take place, as G. Bally said, in a field without tension — in other words, in an atmosphere of relaxation. The urge to explore strives for situations in which the animal can acquire knowledge through its activity. Even when, confronted with an unfamiliar object, it performs the actions proper to a specific activity pattern, such as feeding, it has no desire to eat the object but only to find out whether it is actually edible 'in principle'. The survival value of exploratory behaviour thus lies in the acquisition of objective 'theoretical' knowledge.

Since the behavioural programme of exploratory animals is open to modification to a large extent, their organs and motor patterns need to be used over a correspondingly wide range of activities. Morphologically they are unspecialized, omnivorous; they are frequently also cosmopolitan. Exploratory behaviour, like play, with which it is closely connected, is generally restricted to young animals. That man retains his open-minded curiosity throughout his life is due to the retardation of his development and his partial neoteny, i.e. the persistence of a juvenile state of development.

The objectivating function of exploratory behaviour attains a new level of cognitive processes when it becomes directed at the organism's own body. The moment that an anthropoid discovers that its own hand and the object it grasps are both objects in the material world, its activity of grasping approaches conceiving, and what it knows about the constantly inherent characteristics of the object it grasps begins to become a concept.

Imitation, the subject discussed in section 7.7, is strictly speaking not an independent cognitive process. It presupposes that voluntary movements and their reafferences are at an organism's disposal, and is itself a precondition for the evolution of verbal language communicable by tradition. Anthropoids show only a weak faculty of imitating complex behaviour patterns; it is at its most fully developed in man, and also, remarkably, in birds, but here only in 'song', e.g. a very particular kind of sound utterance. Certain birds have an innate acoustical 'template', with which, by trial and error, they shape their song until sound production matches the inherited acoustical pattern.

In man the act of imitation appears to be initiated by kinaesthetic processes. This purely phenomenological fact is virtually all we know about the physiological mechanism of imitation in humans. The process of trial and error is of very little significance. Both humans and birds have an urge to imitate, and they follow this urge for its own sake, without concern for its purpose. In order to convert what has been seen or heard into motor activity, it is necessary to have a highly complex physiological apparatus such as only emerges in the organic world under the strong selection pressure of a particular survival function. The nature of imitation in humans is clear, but in birds it remains a mystery, especially as they do not use it for communication.

The final section (7.8) dealt with tradition, i.e. the transmission of individually acquired knowledge from one generation to the next. Birds and lower mammals sometimes pass on knowledge of a particular object in this way, while apes can even hand down certain techniques. In all these cases the transmission of knowledge is dependent on the presence of the object or objects in question; only with the evolution of abstract thought and human language does tradition, through the creation of free symbols, become independent of the object. This independence is the prerequisite of the accumulation of supra-individual knowledge and its transmission over long periods, an achievement of which only man is capable.

No one of these eight faculties is the exclusive property of man, but in none of them does he fail to outstrip all other sentient beings. Equally, all of them are indispensable to the functions of abstract thought and verbal communication, but except for imitation in its human form, none of them arose in response to, or under the selection pressure of, this collective function. Each of them serves its particular function to which their original characteristics are tailored, which makes it all the more remarkable that they should be integrated to form a new, higher system separated from all other previously extant living systems by a gulf scarcely less wide than that which divides life from inanimate matter.

Chapter eight

The human mind

8.1 *The uniqueness of man*

For good reasons I devoted the whole of Chapter 2 (pp.29-35) to the phenomenon of *'fulguratio'*, the 'creative flash', or, in a wide sense, the invention by which the integration of two pre-existing and independent systems produces a new one with entirely new and unforeseeable characteristics and functions. It is necessary fully to have understood the nature of this process of creation to appreciate the fundamental revolution of all life, brought about by the coming into existence of the human mind. A large proportion of present-day anthropologists, however, do not have such an understanding, and find themselves divided into two rival camps established on two sets of equally false principles.

The one camp, the so-called reductionists, clings to the fiction of the continuity of the evolutionary process, believing that it can only produce differences of degree. We have seen (p. 42ff.) that every step in evolution causes a change not only of degree but also of essence. Yet Earl W. Count, a characteristic spokesman for the reductionist school, writes:

> The distinction between a kingdom of insects and a society of human beings is not, as has been widely assumed, the distinction between a simple automatism on one hand and a complex social automatism and culturalized social consciousness on the other, but that between a culture with a large instinct-component and

an insignificant learning-component on the one side, and one with a high learning-component on the other.

Elsewhere the same writer stresses — rightly, but in contradiction of his above statement — that the creation of symbols is a specifically human achievement. Both statements miss the point of the essential difference between animals and man.

On the other hand, a complete lack of understanding of the creative process of evolution can develop into a stubborn obstacle to knowledge. Particularly, the lack of insight into the relationship between lower and higher levels of integration, into the way the latter depend on and absorb the former and still are essentially different from them, leads to a complete mix-up of conceptualization. Typological antitheses of mutually exclusive concepts erect almost insurmountable barriers to the understanding of all historical relationships, phylogenetic, cultural and ontological alike. Arguments are invariably based on the antitheses of *man* and *beast* and are expressed in a form which rules out any appreciation of the real historical and ontological relationship between the two. In his Conclusion to Helmut Plessner's *Philosophische Anthropologie*, B. G. Dux writes: 'That man is phylogenetically close to certain of the animals, notably the anthropoids, provides a basic justification for the time-honoured distinction between man and the animals, and shows it to be more than just a recurrent popular motif or a useful stylistic device'. Consequently Dux rejects any conclusions based on analogies between animal and human behaviour, regarding them as products of 'a relatively harmless form of careless thinking'. Philosophical anthropology 'shows that man is a creature who must first create his own world; thus "adaptation" becomes an empty word if what is to be adapted to bears the marks of having itself been designed by man'. For Dux the principle of adaptation is 'an epistemological monstrosity which only survives because it appears to be of some use to ethologists'.

These quotations suffice to show how completely anthropologists of both parties have failed to comprehend the processes of creative evolution, and how inadequate their understanding is of the historical situation. Yet, paradoxically, those who refuse to look at all the things that man and the animals have in common actually underestimate the differences between them. The categorical distinction between man and all other living beings is vital to the purposes of my book, and particularly to those of the present chapter. Nicolai Hartmann called it a 'hiatus' — the great gulf between two different levels of existence, brought about by the 'creative flash' out of which the human mind was born.

In order to avoid confusion between two fundamentally different distinctions, a confusion of which Hartmann himself was well aware, I might interpose at this point a few remarks on that most mysterious of barriers, utterly impenetrable to the human understanding, that runs through the middle of what is the undeniable one-ness of our personality — the barrier that divides our subjective experience from the objective, verifiable physiological events that occur in our body.

Hartmann regarded this 'yawning gulf in the structure of Being' as similar to that which exists 'far below the psycho-physical divide between organic and inorganic nature.' For once I disagree with the great philosopher. These are two basically different distinctions. In the first place his phrase 'far below' conceals a fundamental error. It is not a matter of a horizontal split between subjective experience and physiological events, nor a matter of dividing the higher from the lower, the complex from the more elementary, but a kind of vertical dividing line through our whole nature. There are elementary nervous processes which are accompanied by a powerful sense of experience, and on the other hand highly complex processes which are analogous to rational procedures, yet completely 'non-experienced' and inaccessible to our self-observation. In Hartmann's day the prospect of ever bridging the gulf between the organic and the inorganic by a 'continuum of forms' seemed as remote as that of solving the problem of the relationship between soul and body, so that Hartmann could rightly say: 'The real origin of life, with its self-regulating metabolism and its power of self-reproduction, is something we have not as yet been able to demonstrate.'

But such far-reaching discoveries concerning the constituent functions have since been made in bio-cybernetics and biochemistry that it is now a far from merely utopian hope that in the foreseeable future we shall be in a position to explain the autonomous principles of organic life in terms of its material structure and its evolution. At all events it is in principle far from impossible that the growth of our knowledge will eventually bridge the gap between the organic and the inorganic by means of a 'continuum of forms'.

The gulf between subjective experience and objective-physiological reality, however, is of a different kind, for it is not caused by a gap in our knowledge but by a basic inability *ever* to know, an inability determined *a priori* by the structure of our cognitive apparatus. Paradoxically, the impenetrable barrier between the physical and the spiritual applies only to our reason, not to our feelings. As I said at the beginning of this book (p. 4), when we talk about a person we do not mean just the objectively verifiable facts about his appearance or the psychological facts of his personal experience, but what we unquestioningly accept as the inseparable unity of the two. In other words, despite all rational considerations, we are not able to doubt the reality of this unity. Not without cause did Max Hartmann describe the relationship between soul and body as 'a-logical'.

This book is not concerned with the body-soul problem. What matters here is solely the fact that the gulf between the physical and the spiritual is of a fundamentally different kind from that between the organic and the inorganic, and that between man and the animals. These latter gulfs represent transitions, and each owes its existence to a unique event in the evolution of the world. Not only can both be bridged by a theoretical continuum of intermediate forms, but we also know that these transitional forms have indeed existed at specific moments in time. There are two reasons why the gulf *appears* unbridgeable: one is that, in both cases, the transitional forms were unstable, i.e. they were phases that quickly passed; the other is that the evolutionary step forward was large enough in both cases to open up an immense gap between the two sides of the gulf that had just been bridged.

The 'hiatus' between soul and body, on the other hand, is indeed unbridgeable, albeit perhaps 'only for us', as Nicolai Hartmann put it — that is, with the cognitive apparatus at our disposal. Yet I do not believe that this is a limitation imposed just by the present state of our knowledge, or that even a utopian advance of this knowledge would bring us closer to a solution of the problem. The autonomy of personal experience and its laws cannot in principle be explained in terms of chemical and physical laws or of neuro-physiological structure, however complex.

The other two great gulfs, the one between the organic and the inorganic, and that between man and the animals, can be bridged, because the evolutionary processes involved in both contexts are open to investigation by the methodology of natural science — indeed, they bear a mysterious similarity to each other. The parallels — one is tempted to call them analogies — between these two 'creative flashes', the greatest in the history of our planet, must cause us to ponder deeply. In my first chapter I attempted to show that in one of its vital aspects life is a cognitive process. It came into being with the 'invention' of a structure possessing the power to acquire and retain information, and capable at the same time of amassing out of the streaming world energy around it a sufficient quantity for it to keep the flame of knowledge burning. The sudden creation of this cognitive apparatus produced the first great gulf.

The second great gulf, that between the highest of the animals and man, was produced by another 'creative flash', which, like the first, produced a new cognitive apparatus.

From the virus-like predecessors of living organisms up to our closest animal ancestors the structures and mechanisms for collecting adaptive information remained virtually unchanged. To be sure, individual learning, with the increasing complexity of the central nervous system, came to play an ever more important role, and even the transmission of acquired knowledge from one generation to another began to contribute towards the permanent preservation of such knowledge. But when one compares the quantity and durability of the information acquired through learning and tradition with that stored in the genome, one can only conclude that even in the highest of pre-human anthropoids the division of labour between the genome and those mechanisms. which acquire instantaneous information and store it by learning has remained virtually the same. Even what the most intelligent ape can learn individually and gain from the tradition of its tribe would, if one could quantify it in 'bits' of information, represent only an infinitesimal fraction of the information stored in the genome of its species. Even the knowledge contained in the sequences of nucleotides in the lowest of living organisms would, and indeed does, fill many volumes when expressed in words.

We are, therefore, neglecting only a very small quantity of learned information if we state that during the immense periods of earth history, during which life worked its way upward from creatures far more primitive than present-day bacteria to prehuman primates, it was the chain molecules of the genomes that were charged with the function of gathering, conserving and increasing knowledge. Then, suddenly, towards the end of the tertiary age, a totally different organic system makes its appearance, undertaking to achieve the same purpose more quickly and more efficiently.

If one were to attempt to define Life, one would surely have to include in the definition the function of acquiring and storing information, together with the structural mechanisms of the genome which are performing these tasks. But such a definition would not include the particular characteristics and functions of man. What would be lacking is what constitutes the life of the human mind — and this, one can say without exaggeration, is a new *kind* of life. We must now devote our attention to the nature of this life.

8.2 *The inheritance of acquired characteristics*

Learned behaviour, already extensive and of considerable importance in the most advanced animals, becomes incalculably more important in man. Reflection and conceptual thought make it possible to give permanence to the data provided by mechanisms which were originally used only to acquire short-term information; these data can then be incorporated into the total hoard of accumulated knowledge. Momentary insights are retained, processes of rational objectivation are raised to a higher level of cognition and thus acquire a new significance. Above all, the newly formed tradition of conceptual thought exerts an immense influence on the operation of all learning processes.

By thus becoming independent of concrete objects tradition makes all learned knowledge potentially hereditary. As I mentioned earlier, we are in the habit of associating the word 'inherited' with a specific biological process, overlooking its original legal meaning. When the bow and arrow were invented they became the property not only of the inventor's descendants but of his whole society, and subsequently of all mankind. The likelihood of their falling into desuetude was as remote as that of a physical organ of comparable survival value becoming vestigial. Cumulative tradition signifies no more and no less than the inheritance of acquired characteristics.

The communication and diffusion of human knowledge is so rapid and so marvellous that one might almost be forgiven for overlooking that the foundation of the human mind is physical matter. A flash of individual genius can make an enormous and lasting contribution to the stock of human knowledge, and on occasion a young man's ideas may affect not only the scientific techniques of his senior but even his entire mental outlook. Such was my experience as a young man with my friend Gustav Kramer, who was eight years younger than I. He had grown up in the school of Max Hartmann, a biologist of considerable standing, whose approach to the question of objective reality had been strongly influenced by the philosopher Nicolai Hartmann. I took over this approach from Kramer, as a matter of course, though neither of us was aware at the time of the tradition behind it, and the 'inheritance' of a whole complex of epistemological attitudes crossed the divide between 'generations' and back again.

The transmission of acquired characteristics brings about that acceleration of development, found in all spheres of human life, which may well be one of the causes of the decline of individual human civilizations after a certain period of time. There has been a great deal of discussion since Darwin as to whether acquired characteristics are hereditary or not. Once, half in jest, I coined an aphorism that is appropriate here: 'One is often made aware of the fact that a certain process does not usually occur, by coming across an exception which shows us what things would be like if it did occur.'

8.3 *Intellectual life as a collective phenomenon*

As we have seen, the highly organized social life of the primates provided the basis on which pre-existing cognitive function could be integrated to form conceptual thought and with it syntactic language and cumulative tradition. These faculties in their turn had far-reaching repercussions on man's social life. The rapid spread of knowledge, the assimilation of the opinions of all members of society, and especially the consolidation of certain basic social and ethical attitudes, created a new form of community, an essentially new kind of living, even what we call intellectual life.

The specific, concrete realization of such a system is what we call a culture. It is in my opinion profitless to distinguish between cultural, spiritual or intellectual life in modern man. The new system has all the known constituent properties of the lower forms of life; the positive feedback cycle of information and energy continues to work in basically the same way, even though certain of the physiological and physical processes involved are completely new.

As I pointed out in the preceding section, the transmission of assimilated knowledge in the living system of a human culture is based on mechanisms different from those of the analogous processes in a species of animal or plant. Similarly, the mechanisms for assimilating energy, which rely on such information, are also partly new, for man is the only creature able to exploit for the benefit of his species sources of energy other than the solar energy which feeds the life cycle through the process of photosynthesis.

However, as a result of the new 'inheritability' of acquired characteristics, a new cognitive apparatus emerges, the functions of which are parallel to those of the genome in that the processes of the assimilation and retention of information are performed by two different kinds of mechanism which stand in a mutual relationship both of antagonism and of equilibrium.

8.4 *The social construction of reality*

The tradition of a man's culture prescribes what he learns and how he learns. In particular it lays down limits to what he is allowed to learn. From Berger's and Luckmann's book *The Social Construction of Reality* we can see to what degree our cognitive functions are influenced by what the culture to which we belong considers to be 'real' and 'true'. Our innate perceiving apparatus is overlaid by an intellectual cultural superstructure which, much like innate cognitive mechanisms, provides us with working hypotheses that determine the course of our subsequent individual search for knowledge. This cultural system possesses its own structures which, like all structures, entail a certain restriction of freedom. The information on which these working hypotheses are based, however, does not derive from the store encoded in the genome but from the traditions of our culture, which are far newer and coded in a much more adaptable form. At the same time they are less fully tested, and hence less reliable, though better able to cope with modern demands.

As already explained (p.22), the structures in which all the cultural knowledge determining our 'view of the world', as well as some of our moral attitudes, are laid down, are of course just as material, in fact just as physico-chemical as are the hoards of information contained in the genome, though distinguished from the latter by not consisting of living matter. Thus, much of what a culture has amassed in collective knowledge and what determines the attitudes and world-view of its members is written down, or in modern times recorded on film or tape.

But in spite of such man-made memory banks the central nervous system is required to perform prodigious feats in storing the data of a cultural tradition. Arnold Gehlen, who described man as an 'incompetent creature' and emphasized the inadequate adaptation of his organs, overlooked the fact that the human brain is an organ which is extremely well adapted to the demands of human life. There can be no doubt that the cerebral hemispheres of our ancestors began to enlarge at the moment when the 'creative flash' of conceptual thought and language made acquired characteristics heritable; as a result, selection pressure caused the enlargement of the hemispheres — a view that I have heard Jacques Monod express *en passant* in a debate as though it were a matter of course. I have never come across this plausible proposition in print, though I have read a great number of far-fetched explanations of the sudden enlargement of the cerebrum at the time man emerged. For the creation of man *is* the 'creative flash' of accumulated tradition, and the cerebrum is its organ.

We are barely aware of the immense mass of cultural information contained in our tradition. What we know by tradition tends to become second nature to us, and we naively believe in its reality and its truth in the same way that the naive realist assumes that the empirical evidence provided by his organs of perception constitutes objective reality. As I showed in the Prolegomena, the process of objectivation, which is a prerequisite for all advances in scientific knowledge, is based on an understanding of the subjective mechanism that reflects the images of external reality. Few people realize the extent to which our culturally conditioned world-view apparatus, and hence everything we consider true, certain, real and right, is determined by the social influences of the culture in which we grow up. To the scientist aiming to objectivate the phenomenal world it is a duty to take into account the achievements as well as the limitations of our culturally determined 'world-view apparatus' in exactly the same way as he treats the inherited *a priori* mechanisms of our cognition.

This is much the hardest task facing the man who seeks an objective understanding of the world. In the first place, the presence of value judgements which have become second nature to us creates a serious obstacle — in fact, a number of philosophers repudiate the views expressed in the chapters that follow because they offend their sense of cultural values. And in the second place, intellectual and cultural life constitutes the most highly integrated living system that exists on our planet, so it is hard for us to rise to an even higher level from which to consider it.

Yet this is what we are bound to do. And precisely because substantial parts of our perceiving apparatus derive from our own culture, the demand for objectivity forces us to accept the formulation of Bridgman's that I quoted at the outset (p.3). Indeed, it is our special duty to undertake a scientific investigation of cultural and intellectual life in the face of the malaise and decay that threaten our own civilization at this time.

Chapter nine
Culture as a living system

9.1 *Analogies between phylogenetic and cultural development*

I shall start by applying to human culture the method of approach used by comparative phylogeneticists when investigating any living system. If one compares with an open mind the phylogeny of various species of animals and plants with the history of different cultures, one comes to see them as two kinds of life processes which, though taking place on different levels of integration, are both based on the reciprocal feedback between the acquisition of power and knowledge, as in the case of all life processes.

Let us imagine that a zoologist from Mars arrives on earth, a professional phylogeneticist, well-informed about the social habits of various species of animals, in particular insect colonies, but knowing nothing of the specific functions of the human mind and, in particular, of the accelerated development brought about by the hereditability of acquired attributes. Comparing the way New Yorkers house and dress themselves with that of the Papuans of central New Guinea, he could not but think that these cultural groups belonged to two different species, perhaps even two different genera. And since this Martian would have no notion of the order of magnitude of the time spans involved in evolutionary processes, it might well not strike him, in his investigations, that cultural development and phylogenetic development were two different processes.

The analogies between the two are so close that identical methods of research have been developed independently in different fields. Linguists determine the derivation of contemporary words by the same method of comparing similarities and dissimilarities as that used by comparative morphologists to reconstruct the origins of physical characteristics, and the comparative method can be applied to all fields of cultural development. This discovery has been held back by the same idealistic and typological habits of thought that were responsible for the belated appearance of a true science of comparative ethology, and right down to recent times such ideological considerations continued to lead many philosophers of history to cling to the postulate of a single, unitary development embracing the history of all mankind. Spengler, Toynbee and others, building on Herder over a century earlier, showed that the unity of human civilization is as much a fiction as is the unity of the phyletic development of the tree of life. As every branch and every species develops at its own risk and in its own direction, so also does every individual civilization. As Hans Freyer points out, an area which had hitherto known only rural settlements and loose tribal associations can apparently give birth overnight to temples and pyramids, fortified cities and state authorities.

That this phenomenon has been observed in various places and at various times in the history of the world has seduced philosophers of history, as Freyer puts it,

> ... to construct a kind of model which, while not explaining the miracle of high culture, at least reduced it to an intelligible formula. But the further modern historiography penetrates into the distant past, calling on the techniques of archaeology and comparative linguistics, the more questionable these models become, and the more clearly it emerges that the beginnings of culture are as individualized as its subsequent development.

Cultures do not emerge in linear succession or follow a uniform pattern of laws, as postulated by the 'unitary' philosophy of history, but develop independently of each other, just as do the different species of animal and plant — a polyphyletic development, as a phylogeneticist would call it.

Thus the complex living systems that we, as well as the historians, call high cultures probably arose as a result of 'creative flashes' analogous to the advances in the evolutionary process that have produced the different species of animals. One might even put forward the notion that every advance in culture is due to the integration of pre-existent and previously independent subsystems. High cultures often, though not always, seem to arise when a people moves from its native area and comes into close contact with another people that has remained sedentary, and a 'graft' of foreign culture — 'la greffe' as Paul Valéry called it — frequently appears to have stimulated 'creative flashes'. This, of course, is a possibility only given to cultural development, as it presupposes the inheritance of acquired characteristics.

The influence that one high culture can exert on another by 'grafts' of this kind may well be considerable, yet the differences between one culture and another are so extraordinary that even to call them fabulous is an understatement.

If one compares phenomena of a similar order of magnitude — a hieroglyphic inscription and a cuneiform text, the pyramids of Gizeh and the ruins of the Ziggurats of Erech and Ur, a relief from the Old Kingdom and the contemporary stele of victory of King Naram-sin of Akkad, the Minoan palaces of Crete, together with their murals, and the architecture and painting of the Hittite kingdom of the same period — one finds oneself plunged from one world into another, and it requires a considerable intellectual effort to prevent the differences from turning into opposites. (Freyer)

This profound statement is as true of the gamut of manifold living forms produced by evolution as it is of the multiplicity of cultures produced by human history. It does indeed require an intellectual effort to overcome our innate tendency to range the elements of the phenomenal world in pairs of antitheses, and many thinkers are prone to work in terms of disjunctive, mutually exclusive concepts. The *a priori* compulsion to think in opposites may be important as a primal means to order concepts, but it is very necessary to hold it in check: particularly so when the issue is that of grasping the historical significance of a multiplicity of forms that have evolved as a genealogical tree — that of species as well as that of cultures — through growth and diversification. Thinking in opposites is at the root of typological systematics and it seems to be particularly deep-seated among the Germans: at least it has prevented many great thinkers, Goethe among them, from discovering the phylogenetic evolution of living things. The failure to make the 'intellectual effort' of which Freyer speaks, is as great an obstacle to the advancement of knowledge in the realm of cultural history as it is in phylogeny.

Among the ideological obstacles to accepting the facts of cultural history as they really are is the idealistic belief that there is an overall plan underlying the whole of history. That history, like phylogeny, is governed by chance and necessity is a fact that, as experience has shown, a great number of people find impossible to accept.

9.2 *The phylogenetic foundations of cultural development*

The preceding discussion of the parallels and analogies between phylogenetic and cultural evolution might lead one to imagine that we are dealing with two processes which are interchangeable but yet quite independent in origin and each with its own separate course to run. But this would once more leave the door open for the misleading tendency to think disjunctively in opposites. The same tendency is responsible for the widespread view that cultural development is somehow sharply divided from the phylogenetic development which preceded it and which is considered as having come to a stop with the 'creation' of men.

The same false assumption underlies the attitude that all the 'higher' things in human life, especially the refinements of social conduct, are the results of 'culture', whereas all the 'lower' things are the products of instinct; in fact, it is a typical phylogenetic process that has made man the civilized creature that he is. The radical change brought about in the human brain by the selection pressure of the accumulation of transmitted knowledge is not a cultural, but a phylogenetic process, and it took place *after* the 'creative flash' in which conceptual thought was born. The adoption of a fully upright posture and an increased differentiation in the muscles of the hand and fingers probably took place at the same time.

Nor is there any reason to believe that the phylogenetic development of our species has come to a halt. Abrupt changes in our environment and the demands it makes upon us suggest that *Homo sapiens* is presently in a state of rapid genetic evolution. This is confirmed by his marked increase in height and by various other symptoms of man's self-domestication. We must accept the fact that there are two kinds of development in man which take place at different speeds but which constantly interact: one, the slower, is that of cultural development.

One of the most important tasks facing ethologists is to differentiate between the effects of these two processes and to trace them to their respective causes. To distinguish norms of social conduct that are phylogenetically programmed from those that are culturally determined is of great practical importance, because completely different therapeutic measures are indicated in pathological disorders according to whether they are attributable to the one source or the other. It is also of fundamental importance for theoretical purposes to establish the origin of the adaptive information on which the survival value of a particular behaviour pattern is based.

The comparative method provides us with a number of means for carrying out the necessary analysis. One is to determine the relative speed with which a particular feature or group of features is modified in the course of time. Even before the emergence of the hereditability of acquired attributes and the rapid increase in their rate of modification, the phylogenetical modification proceeded at rates that were different for various components and structural principles. The cell nucleus, for example, has remained unchanged from protozoa to man, and the microstructure of the genome is older still. On the other hand, the macroscopic structure of the various phyla of the animal kingdom has assumed all manner of forms over the same period of time. A toadstool and a lobster, an oak tree and a human being are so different from each other that, if we knew only these 'top branches' of the tree of life, it would not readily occur to us that they had sprung from a common root — which they undoubtedly did. It is this bewildering diversity of forms that tempts us into the error of desperately trying to classify things into cut-and-dried types and groups.

Of central importance in comparative ethology is the determination of the relative rate at which different characteristics or groups are changing. Those common to large groups can appropriately be called 'conservative'. As the structural principle of the nucleus has the same form in all nuclear organisms, one can properly conclude that it is a very old structure with great 'taxonomic rank'. Conversely, the smaller the taxonomic group united by a particular characteristic, the more recent the characteristic in question. There are all manner of intermediate stages between the slowest and fastest forms of phylogenetic change. The speed of the fastest approximates to that of cultural processes; thus many domestic animals have changed so radically since the time of their wild ancestors that one is entitled to regard them as new species.

Yet the speed of this most rapid of phylogenetic processes known to us is so much slower than that of average cultural change that the difference can help us to distinguish the two kinds of phenomena. If we discover that certain behaviour patterns and norms of social conduct are found in all human beings of all cultures in exactly the same form, we can assume with virtual certainty that they are phylogenetically programmed and genetically specified — for it is manifestly improbable that such norms could remain unchanged over so long a period if they were merely the product of tradition: in other words, we know that they are inherited from one and the same ancestral culture.

A surprising double confirmation of this comes from two branches of science which appear at first sight quite unrelated to each other. The first is that of comparative human ethology. There are good reasons for believing that our emotions, which are basic to the motivation of our social conduct, contain a large number of inherited, phylogenetically fixed factors. As Darwin already knew, the ways in which we express our emotions involve a large number of innate, specifically human behaviour patterns. Starting from this point, Eibl-Eibesfeldt made a comparative study of the way in which feelings are expressed, encompassing as many independent human cultures as were accessible to him. He filmed a selection of gestures and expressions that could predictably be elicited by certain standardized situations, such as greeting, bidding farewell, quarrels, fear, happiness, terror and so on. The camera had a prism set in front of the lens so that it photographed at right angles to the direction in which it was ostensibly pointed, leaving the subjects being filmed to behave freely and naturally. The result was as simple as it was surprising. For after the closest of analyses, even in slow motion, the behaviour patterns of Papuans from New Guinea, the Waika Indians of the Upper Orinoco, the Bushmen of the Kalahari, Australian aborigines, cultured Frenchmen, South Americans, and other representatives of our Western culture, all proved to be identical.

The second line of research which, proceeding quite independently, arrived at amazingly similar results, is linguistics — that is the comparative study of languages and their inherent logic. In a discussion of the problems of communication, my wife once voiced her surprise that language can be translated at all. Nevertheless, anybody trying to learn a language will confidently ask what are the equivalent terms for 'though', 'already', 'albeit' and so on in Latin, Hungarian or Japanese, and will be rather surprised if, exceptionally, there is no strictly corresponding word. That, as a rule, such an equivalent does indeed exist can by no means be taken for granted. However, comparative linguistics has taught us that men of all races and all cultures possess certain thought patterns which are innate, and that not only are these patterns at the basis of the logical structure of language but also that they control the logic of thought itself. Chomsky and Lenneberg reached these conclusions through a comparative study of the structure of language. Höpp followed a different path to arrive at similar views on the unity of thought and language, showing, as he put it in his *Evolution der Sprache und Vernunft* (1970), 'how mistaken it is to divide the mind into an outer part, i.e. language, and an inner part, i.e. thought, when both are in reality two aspects of one and the same thing'. The only philosopher ever to voice similar opinions is the Austrian, F. Decker.

No one will deny that in the course of their evolution language and conceptual thought exercised a mutual influence on each other as they became increasingly sophisticated. We can see from self-observation how we have recourse to linguistic formulations when pursuing complicated trains of thought, even if only for mnemonic purposes, in much the same way as we use pencil and paper to make calculations. There is no doubt that the structures of logical thought were present before syntactical language, but there is not any doubt either that they would never have reached their present level of differentiation if this interaction of thought and speech had not occurred.

Apart from the comparison of related forms, i.e. those with a common ancestry, ethologists have another means of distinguishing between behaviour patterns that have been individually acquired and handed down, and innate patterns of phylogenetic origin. This is to raise an animal from birth under artificial conditions, one of these being to deliberately deprive it of the means of acquiring certain information. This is obviously not an experiment that can be performed on human beings, but it is possible to assess the analogous situation in which children are born deaf and blind. Using the same principles and methods as those described above, Eibl-Eibesfeldt filmed and analysed the emotional expressions and actions of these unfortunate children. The results showed that those very expressions which Eibl-Eibesfeldt has found to be identical in representatives of the most diverse cultures now reappeared almost without exception in these children. This disposes of the theory, still stubbornly defended by many anthropologists, that all human communication and social behaviour is determined exclusively by cultural tradition.

Chomsky and his school reached their conclusions, as mentioned above, on the basis of the same principles as those followed by Eibl-Eibesfeldt in his demonstration of the genetic programming of emotional expressions — namely, by seeking to deduce the laws prevailing in all human cultures alike. The conclusions of both men receive considerable support from the results of ontogenetic research, especially that carried out in conditions of specific deprivation.

When, in normal circumstances, a child starts to talk, it is clear that he does not imitate words and phrases like a parrot but possesses from the outset certain rules of sentence construction. As Otto Koehler once put it, a child does not actually learn to talk but only learns words. There is seldom much profit to be derived for the analysis of innate structures of thought and language from the study of children born blind and deaf, simply because it very rarely happens that a lesion causing loss of sight and hearing does not at the same time damage the brain to such an extent that logical thought is seriously impaired.

We do know, however, of one such case that is of very great importance, an importance often underestimated these days because of the idiotic fashion of refusing to regard as a legitimate source of scientific knowledge any isolated event or observation that cannot be repeated or reduced to statistical terms. I am referring to Anne M. Sullivan's simple account of the mental development of the blind and deaf mute, Helen Keller. The value of this document cannot be too heavily stressed. For here is the story of how, through a coincidence of uniquely favourable circumstances, a talented teacher, endowed with great powers of observation, came into contact with a highly gifted, almost inspired child whom nature had condemned to be without sight or hearing.

It is worth asking why the valuable information that this remarkable story contains is not better known among psychologists and ethologists. I think I have the answer. Anne Sullivan's account of how quickly her pupil mastered the apparently insuperable task of learning language solely on the basis of feeling the imprint of the letters of the alphabet on her hand, and how she came to form abstract concepts of the most complex kind — all this must strike anyone biased in favour of behaviourist views on learning as utterly incredible. If one knows something about ethology and the above-mentioned research in comparative linguistics, however, one can accept Anne Sullivan's story without hesitation, although one may also be surprised by more things than she herself appears to have been.

Helen Keller was born in Alabama on June 27, 1880, and Anne Sullivan began to teach her on 6 March 1887. Up till then Helen had spent almost all her time sitting on her mother's knee, seeking comfort in physical contact. She had two gestures, which she used to show that she was hungry or thirsty, but could not understand symbolic or linguistic communication of any kind. Miss Sullivan began by using the finger alphabet to write not only words but whole phrases on Helen's hand, just as one teaches a normal child by talking to it. Two days later she gave Helen a doll — a toy she appeared to have played with before — and spelt the word 'doll' on the palm of her hand. She then continued this procedure with other objects, invariably using the standard letters of the alphabet, not simplified symbols.

If, before reading about this, I had been asked whether a blind and deaf person could learn to read in this manner without first having learnt to talk, I would without hesitation have answered no. But from the very first day Helen not only formed a mental link between the signal and actually receiving the object in question but also — which is far more — reproduced and returned the signal. She had not yet, of course, reached the point of turning the letters she had felt into a concept, and was only reacting to a group of tactile impulses, reproducing them in an imperfect, though recognizable form. But it is almost beyond the bounds of credibility that she should even have attempted to do such a thing.

Helen had a dog, of which she was very fond. On 20 March she tried to communicate with it by writing the first word she had learnt, 'doll', on its paw. By 31 March she knew eighteen nouns and three verbs, and had begun to ask the names of things by taking them to her teacher and holding out her hand for her to write on. She clearly felt a need to make mental links of this kind. It reminds one of the story of Adam, who was told to name the things that God brought him.

That Helen had not yet fully grasped the principle of the symbolism of language emerged *inter alia* from the fact that to start with she could not distinguish between nouns and verbs. 'The words "m-u-g" and "m-i-l-k", wrote Anne Sullivan, 'have given Helen more trouble than other words. When she spells "milk", she points to the mug, and when she spells "mug", she makes the sign for pouring or drinking, which shows that she has confused the words. She has no idea yet that everything has a name.'

This introduces us to a very important matter which Anne Sullivan also mentions earlier in her story, although even more casually than here. Helen imitated certain actions in order to make herself understood. This amounts to a form of symbolism, as do the signals she had already learnt to understand and even to transmit. Initially all these signals were associated with actions of some kind — 'doll' with play, 'cake' with eat, and so on — so that she could not distinguish at first between 'mug', 'milk' and 'drink'.

The decisive development of separating symbols from objects and actions came about in so dramatic a way that it is best described in Anne Sullivan's own words:

This morning, while she was washing, she wanted to know the name for 'water'. When she wants to know the name of anything, she points to it and pats my hand. I spelled 'w-a-t-e-r' and thought no more about it until after breakfast. Then it occurred to me that with the help of this new word I might succeed in straightening out the 'mug-milk' difficulty. We went out to the pump-house, and I made Helen hold her mug under the spout while I pumped. As the cold water gushed forth, filling the mug, I spelled 'w-a-t-e-r' in Helen's free hand. The word coming so soon upon the sensation of cold water rushing over her hand seemed to startle her. She dropped the mug and stood as one transfixed. A new light came into her face. She spelled 'water' several times. Then she dropped on the ground and asked for its name and pointed to the pump and trellis, and suddenly turning round she asked for my name. I spelled 'teacher'. Just then the nurse brought Helen's little sister into the pump-house, and Helen spelled 'baby' and pointed to the nurse. All the way back to the house she was highly excited, and learned the name of every object she touched, so that in a few hours she had added thirty new words to her vocabulary. Here are some of them: *door, open, shut, give, go, come.*

The next day Anne Sullivan added:

Helen got up this morning like a radiant fairy. She has flitted from object to object, asking the name of everything and kissing me for very gladness. Last night when I got in bed, she stole into my arms of her own accord and kissed me for the first time, and I thought my heart would burst, so full was it of joy.

As we know from the earlier part of Anne Sullivan's account, up to this point the relationship between them had been that of teacher to pupil, a relationship of authority established only after a long struggle, sometimes involving physical discipline. From now on, however, respect was joined by affection and gratitude.

It is as typical as it is regrettable that in recent literature on deaf-and-dumb children there is not a single account to be found in which one can follow with anything like the same clarity as in this underestimated book how a mind can come to understand the significance of symbols. Anne Sullivan was an intelligent, almost inspired woman who had the rare fortune to be allowed to educate an exceptionally gifted child deprived only of its sensory faculties, with no impairment of the other functions of the brain. It was also a stroke of fortune that the struggle between teacher and pupil took place as early as it did. Any later, and her findings would have been demolished by the objections of that school of psychologists which would dub all her descriptions anecdotes and regard all emotional involvement as anti-scientific. In her naiveté and the warmth of her heart, Anne Sullivan did what was educationally correct and drew the right ethological conclusions from her results. Some of her observations on the way children learn to talk come very close to our notions of so-called 'innate learning dispositions' as investigated by Eibl-Eibesfeldt, Garcia and others. Thus she says: 'A child is born with the capacity to learn, and will learn by itself, provided the necessary external stimuli are present.' And elsewhere, referring to Helen: 'She learns because she cannot help it, just as a bird learns to fly.'

Chomsky's investigations have shown how discriminating, how finely and comprehensively controlled, are the innate human mechanisms of abstract thought.[5] We do not learn to think, we learn the symbols for things, like a vocabulary, and the relationships between them. What we have learnt we then set into a preformed framework without which we would be unable to think

— indeed, without which we would not be human beings at all. But there are hardly any circumstances that illustrate the presence of these various mechanisms so vividly as the simple and completely unbiased description for which science is forever deeply indebted to Anne Sullivan.

Above all, the astonishing speed with which Helen Keller developed a faculty for conceptual thought shows that it was not a question of providing something that was missing but of activating something already present. On the very day her education started, she reproduced the symbols shown to her. Two weeks later she tried to spell to her dog. Eleven days after that she knew over thirty words, four of which were the result of direct asking; after another five days she suddenly grasped the distinction between nouns and verbs, and realized that every object and every activity had a name. A further nineteen days later she began to construe sentences. When she wanted to give her newly born sister a hard sweet and was prevented from doing so, she spelled out: 'Baby eat — no'. Then: 'Baby teeth — no, baby eat — no'. Two weeks afterwards she could use the conjunction 'and'; when she was told to shut the door, she added spontaneously 'and lock'. The same day she found that her dog had a litter of puppies, and she spelled out to the others in the house, who knew nothing about it, 'Baby dog'. The same day, after fondling and stroking the puppies, she made the remark 'eyes — shut, sleep — no'. The same day she learned to use the word 'very'.

Her urge to learn words with new connotations was shown by the fact that when she began to grasp the meaning of a new word she regularly put her understanding to the test by using the word in a variety of ways. With 'very', for instance, she said: 'Baby small, puppy very small' (she had learned at once to say 'puppy' instead of 'baby dog'). Then she produced two small stones of different sizes, showed them in succession to her teacher and said: 'Stone small — stone very small'.

Less than three months after the time when she had still been unable to communicate a single word, Helen Keller wrote in Braille a very coherent letter to a friend, and was so keen on reading that she used to smuggle a book in Braille into her bed at night and read beneath the bedclothes. When she was shown a litter of new-born piglets a few days later, she asked: 'Did baby pig grow in egg? Where are many shells?'

By the end of July she had learnt to write legibly and quickly with a pencil. At this period she also discovered the questions 'Why?' and 'What for?' and her eagerness to learn became almost irksome. In September she started to use pronouns correctly, and soon afterwards the verb 'to be' made its appearance. For a long time she considered the article unnecessary. A year later she had mastered the subjunctive, using it correctly and with special pleasure in contexts of wishful thinking and in the conditional, and all in all cultivated so refined and elegant a style that it appeared almost affected in an eight-year-old girl.

We must never forget that Helen's entire experience, including that of what is beautiful and what is good, came through the letters of words written on her palm — that is, through language. Small wonder that she loved it inordinately. As Anne Sullivan says, 'Helen acquired language by practice and habit rather than by study of rules and definition. Grammar, with its puzzling array of classifications, nomenclatures and paradigms, was wholly discarded in her education.' Miss Sullivan never communicated with Helen in simplified sentences but always used a correct syntax, without omitting adverbs, prepositions or any other element. Helen herself, however, did communicate at first in rudimentary sentences, adding further elements one by one, the last being the article. Anne Sullivan's comment on the extreme rapidity with which Helen mastered this task was: 'It seems strange that people should marvel at what is really so simple; why, it is as easy to teach the name of an idea, if it is clearly formulated in the child's mind, as to teach the name of an object. It would indeed be a herculean task to teach the words if the ideas did not already exist in the child's mind.'

This overlooks the fact, however, that it would have been an incomparably greater achievement than learning the meaning of symbols and acquiring a vocabulary, if Helen had been able fully to conceptualize all the complicated rules of grammar and linguistic logic between March 1887 and September 1888. The manifest impossibility of such a mammoth achievement is for me an unshakable proof of the correctness of Chomsky's theories.

The innate structures of thought and language, as well as the universal forms of emotional expression demonstrated by Eibl-Eibesfeldt, are but two examples of behaviour patterns which our species has evolved in the course of its history and preserved in its genome. There must be numerous other such patterns. As I mentioned earlier, Darwin had already surmised that most emotions, and the behaviour patterns pertaining to them, were 'instinctive', i.e. the genetically specified property of our species. Eibl-Eibesfeldt's specific researches have made it very much more probable that Darwin's general assumption was correct. Innate norms of behaviour must play a particularly important part in the structure of human society. In their book *The Imperial Animal* two anthropologists, Lionel Tiger and Robin Fox, have successfully attempted to employ a methodology similar to that of Eibl-Eibesfeldt and Chomsky in order to produce a 'bio-grammar' of human social behaviour. It is an inspired undertaking, and, for all their boldness, its basic conclusions are thoroughly convincing.

All inherited behaviour programmes share a resistance to the modifying influences of human culture. The highly differentiated complexes of behaviour patterns which have been discovered recently by so very different independent investigations are common to all men and all cultures and are fixed and invariable in form. The great significance of this fact is most readily assessed by those who know, like Hans Freyer (p.178 f.), how divergent the customs, ideals and achievements of different cultures are.

The identical nature of such behavioural norms indicates far more than just their independence of cultural influences; it also demonstrates a fundamental and insuperable resistance to them. This suggests in its turn that they form a framework, a sort of skeleton, for our social, cultural and intellectual behaviour, a kind of blueprint of the form of our society. As Arnold Gehlen says, man is by nature a civilized being — that is to say, his natural, inherited character is so constituted that many of its structures cannot fulfill their function if they are divorced from cultural tradition. On the other hand, it is they that make the existence of tradition and culture possible, but without a hoard of culturally accumulated knowledge the frontal cortex, which only evolved as a memory bank for cultural tradition, would remain empty. The same applies to the most important of its parts, the language area, without which there would be no logical or abstract thought, but which, on the other hand, could not function unless the cultural tradition provided it with a vocabulary developed over thousands of years of history.

When one seeks to understand, or to teach others, the structure and function of a complex living system, one normally begins with those aspects that are least susceptible to modification. Every anatomy textbook starts with a description of the skeleton. This way of doing things is based, reasonably enough, on the principle that, when one studies the numerous interactions that constitute a living system, the least variable factors are most frequently encountered as causes, and least frequently as effects.

In ethology it has proven good strategy to begin studying a particular animal by making an 'ethogram', i.e. an inventory of the phylogenetically programmed behaviour patterns proper to the species in question. One obstacle to a better understanding of human behaviour is the persistent, stubborn, doctrinaire refusal on the part of philosophical anthropologists even to consider the possibility that there might be genetically programmed behaviour patterns in man. This is the more to be regretted, since hereditary invariants in human behaviour are undoubtedly of great importance in the pathology of cultural development. The regular decline of high cultures, for example, which has preoccupied so many historians from the Greeks to Spengler, may well have been the consequence of a discrepancy between the rates of development of phylogenetically controlled behaviour patterns on the one hand and tradition-controlled patterns on the other. Cultural development is simply too fast for human nature to keep up with and, as Klages noted, the mind can become the enemy of the soul. We urgently need to know more about these things.

9.3 *Divergent development of species and cultures*

As I showed in the first section of this chapter, the history of different cultures, and the historical relationships between them, can be studied by the same methods as phylogeny and the relationships between animal species. This fact alone proves that there are extensive parallels between the two. And as the way in which a species or a culture emerges is of central importance to our understanding of the concepts behind these terms, we must examine this question more closely.

If a population of conspecific creatures is permanently divided in two — say, if an earth movement were to throw up an impassable barrier across the middle of their area of distribution — the two halves will develop in different directions, even if conditions in the two areas do not cause different forces of natural selection to operate. The fortuitous nature of the processes that cause genetic change is in itself sufficient to cause the two populations, which have no genetic exchange with each other, to develop in their own ways.

Conversely, the constant interchange of hereditary factors prevents a population from splitting into two species, even though its various groups are exposed to different forces of natural selection. As an example, let us imagine that the climate along the northern flank of an area inhabited by a certain species of mammal causes the animal to develop a thicker fur, while along the southern flank contrary conditions prevail. As long as there is unrestricted genetic interchange between the animals of north and south, the morphological development of the species has to strike a compromise between the two pressures; if there is no such interchange, the relevant selection pressure may breed the genetic factors required for the particular adaptation. In Ernst Mayr's well-considered opinion, the formation of an entirely new species can only take place in conditions of isolation, generally geographical. In the case of widely distributed species long distances also tend to have an isolating effect, and populations living in widely split-off margin areas can be genetically very different from each other. In this case they are usually defined as races or subspecies.

Since there can be any number of intermediate stages between unrestricted genetic interchange and its complete cessation, the decision is often left to the taxonomist whether to regard two types of animal as races, subspecies, or 'good' species. The best definition of a 'good' species is that it represents a 'panmictic' population, that is one which is unified by genuinely unrestricted interchange of genes. An important criterion for regarding animal varieties as belonging to different species is their ability to exist side by side in the same distribution area without interbreeding.

But there are cases in which all the criteria for defining the concept of species prove inadequate. An instructive example is that of the ancestor of the lesser black-backed gull and the herring gull (*Larus fuscus* and *Larus argentatus*). From an area which we cannot precisely locate, the ancestors of both 'species' spread throughout the temperate zone all round the whole northern hemisphere. As they moved eastwards, they became increasingly darker and smaller — in fact, more like the lesser black-backed gull; as they moved westwards, they became lighter and larger, i.e. more like the herring gull. Over the whole area all possible gradations are found, and unrestricted genetic interchange between neighbouring populations clearly still prevails today. Within the longitudinal limits of Europe, however, black-backed gull and herring gull exist as two 'good' species which normally do not interbreed, although experiments performed in captivity have shown that one can produce fertile hybrids by crossing the two. In other words, the further eastwards and westwards the ancestors of these gulls penetrated, the more different from each other they became, and by the time they had covered the whole globe, they were sufficiently differentiated to be able to overlap in European waters without interbreeding.

When one attempts to classify cultures in terms of degrees of diversity, and to define, for instance, the concept 'subculture', one encounters difficulties very similar to those of the biological taxonomist. There are innumerable subdivisions between large-scale, clearly identifiable high cultures and small communities which differ only slightly from each other, and anyone with a sense of history who does not confuse differences with typological opposites will immediately realize that they are the products of a diverging development. That is the reason why the same methods as in phylogenetics are applicable.

At the same time one must take account of particular features of cultural development which have no place in phylogeny. Firstly, convergent development, especially in the field of technical invention, occurs far more frequently in cultural history than in phylogeny. Earlier ethnologists ignored this fact and regarded all similarities as homologous, which frequently led them to false conclusions. Frobenius's *'Kulturkreislehre'*, dominant in its time, had no room for convergences. Secondly, as mentioned above (p.172 f..), the 'hereditability' of acquired characteristics makes it possible for whole complexes of attributes to be transplanted or 'grafted' from one culture to another. Thirdly, cultures can merge to form a new, homogeneous entity even if each of them has been developing independently over a long period. In other words, it is easier to hybridize cultures than species.

But in spite of these dissimilarities, which are due to basic differences in the processes involved, there is a remarkable resemblance between the emergence of species and that of independent cultures.[6] Erik Erikson, who was, I believe, the first to point out these parallels, coined the term 'pseudo-speciation' for the divergent development of different cultures from a common root. In certain respects cultures that have grown a considerable distance apart do indeed react to each other in a similar manner to that of different, but closely related, species of animals. It is important to stress this closeness of relationship because we know of no case in which the development of two cultural groups, ethologically and ecologically, has been so divergent that they would be able to live independently and peacefully side by side in the same area without trying to compete with each other — as, for instance, teals, shovellers and mallards can without difficulty.

In view of the potential miscegenation of cultures, one cannot but wonder how it is that they have managed, and still manage, to keep themselves 'pure' to the extent they did and still do. In Chapter 10.6 below we shall deal more fully with the subject of how even in the smallest subcultures and ethnic groups individual traditional features of behaviour become status symbols. The customs and manners of one's own group are considered 'refined', those of all other groups, including any rival group with objectively equal status, as 'crude', in a degree directly correlated to the order of the differences. The emotional quality that comes to be attached in this way to the rituals of one's own group, and the complementary denigration of all customs found only in other groups, not only strengthens cohesion within the group but contributes to its isolation from all others, and thus to the independence of its subsequent development. This has consequences analogous to those that geographical isolation has for the evolution of species.

The barriers that these various processes erect between divergent cultural groups are characteristic of all cultures, and are obviously indispensable to their further development. Within animal species intraspecific rivalry inevitably leads to a form of natural selection which, far from being of benefit to the species in its confrontation with its environment, often causes serious harm. Competing against each other with the help of specialized morphological structures, the rivals urge each other on in an ever more intense development of these characteristics, and the cycle often ceases only when grotesque, exaggerated forms find themselves in conflict with other factors of natural selection. The only occasions when antlers are of use to a stag, or wings to an Argus pheasant, are when the animal is fighting its rivals. Yet so strong is the selection pressure behind these characteristics that an animal which lacked them would have no prospect of ever helping to perpetuate the species.

Within a particular culture human competition has similarly adverse effects. However, the tendency of human cultures to split up makes the several cultures almost as different from each other as animal species are. Hence, competition *between* cultures not only does not produce the dangerous effects of other kinds of intraspecific rivalry, but has proved essential to the advance of mankind. It has caused various cultures to compete with each other in various different fields and with various different means. These cultures lived on different foods, used different tools and fought with different weapons. This intercultural rivalry has been one of the most important factors in breeding a higher human intelligence, intellectual subtlety, inventiveness and so on; indeed, these factors were probably responsible at an early stage for the rapid enlargement of the cerebrum, however scornfully philosophical anthropologists may talk about the principle of adaptation as an 'epistemological monstrosity'.

The general direction of organic development, tending from the lower to the higher, is governed by the multiformity of selection pressure, the diversity of the demands made on the organism. If this diversity gives way to an exaggerated and one-sided selection pressure, as in the above example of intraspecific rivalry in animals, evolution will be diverted from the path that leads to new and higher things. Mankind is, at this very time, exposed to just such a pressure, which in many ways is similar to intraspecific rivalry in animals. The dividing lines between cultures are becoming faint or disappearing altogether, and ethnic groups throughout the world are on the point of coalescing to form a single, comprehensive culture embracing all of mankind. At first sight this might appear a desirable development, since it helps to reduce the hatred between nations. But egalitarianism of all peoples also has a destructive effect. For as men of all cultures now fight with the same weapons, compete with each other through the same technology and seek to outwit each other in the same world markets, the force of intercultural selection turns into intracultural selection, whose effects are quite as pernicious as those of intraspecific selection. In a second volume I intend to discuss, among other things, the retrograde evolution of mankind and its culture resulting from the termination of this process of creative selection.

The natural tendency for human cultures to divide and develop in rival directions has dangerous consequences as well as the beneficial effects that have already been mentioned. Prominent on the debit side are hatred and war. As I have described in

On Aggression, the factors that hold small cultural groups together and isolate them from other groups ultimately lead to dissension and bloodshed. The mechanisms of social conduct that originally appear so fruitful, such as an ingrained pride in one's own tradition and a corresponding scorn for the traditions of others, can provoke the most dangerous form of collective hatred, as the rival groups grow larger and as traditions become more uncompromisingly fixed, and their enmity more intense. From the harmless fights which the boys of the Schotten-Gymnasium used to have with the despised 'common' students of the Wasa-Gymnasium, there are all gradations up to the religious wars between nations and ideologies encompassing large portions of the globe, wars in which all the forces of collective aggression are let loose and all inhibitions to killing one's fellowmen removed.

Chapter ten

Factors making for the invariance of culture

10.1 *Equilibrium as prerequisite for further development*

We are able to identify an animal as belonging to a particular species because its gene pool determines a sufficient number of constant characteristics shared by that particular animal population. We call such characteristics 'specific', and it is this gene pool that constitutes the essence of a species.

In the same way as a zoologist recognizes the species to which an animal belongs, so an archaeologist or a historian can see at a glance to which culture, and which period in that culture, an object may be assigned. In view of the ease with which the products of the human mind can be changed by the acquired characteristics that it has inherited, it is necessary to discuss the comparative invariance of cultural processes which permits an expert to make confident judgements of this kind.

The continued life of a species depends on an equilibrium between the invariance of its hereditary factors and their variability. Geneticists and phylogeneticists understand fairly well how an animal or plant succeeds in adapting itself to the changes, some big, some small, that are constantly taking place in its environment. The balance between the factors that make for the invariance of the gene pool and those that make for its modification varies from species to species and is adjusted to the degree of variability of the environment. In relatively stable environments, like the ocean, it is the factors favouring invariance that predominate, while mutation rate and heterosis are at their highest in creatures that inhabit highly variable environments.

The many parallels between the evolution of species and the historical development of cultures suggest that we should also search human culture for two opposed but complementary principles, the interaction of which produces and sustains that equilibrium between invariance and adaptability on which the viability of any living system depends. At this point I cannot avoid anticipating certain aspects of the second volume of the present work, in which I intend to discuss how this balance becomes disturbed when one or other of the counteracting mechanisms fails to function properly, for the little we know about these processes comes mainly from the study of such phenomena of imbalance and malfunction.

To justify my proceeding in this admittedly questionable manner, I may mention that most textbooks of physiology do exactly the same: they describe the normal, 'healthy' process first, notwithstanding the fact that nearly all that is known about it stems from the investigation of its disturbances. The really legitimate way of teaching would be to lead the student exactly along the same path that research has trodden, in its time, however difficult and tortuous this path may be.

The two effects of any structure, that of supporting and that of sacrificing degrees of freedom, confront all living systems, be they organisms, species or cultures, with the same problems, the same necessity of finding a compromise between the two. The earthworm can bend its body in any direction it wants; we can only do so where we are provided with joints, but we can stand upright and the worm cannot. The invariant structures of a species are what make up its adaptedness, and they stand in an interesting relationship to knowledge. Knowledge cannot be stored in any other form than in structure, whether this be the chain molecules of the ganglion cells of the brain, or the letters in a textbook. Structure is adaptation in its finished form. But if further adaptation is to take place and fresh knowledge is to be acquired, a structure must be dismantled and rebuilt, at least in part.

A good example is how a bone grows. On the one hand there are boneforming cells, osteoblasts, which produce layers of new bone tissue that soon becomes sclerotic; there must, therefore, be other cells present, the osteoclasts, capable of destroying the old tissue. The combined activity of the two opposing forces enables the bone to maintain the same proportions and to adapt constantly to the size of the growing animal, so that the size of the bone is in harmony, at every stage of its growth, with the organism as a whole.

All accumulation of human knowledge as a necessary constituent of cultural being depends on the creation of firm structures. These structures need to possess a relatively high degree of invariance in order to become inheritable and to be passed down cumulatively over substantial periods of time. The sum total of information possessed by a culture, residing in its habits and customs, its methods of agriculture and its technology, in the vocabulary and grammar of its language, and above all in its conscious, learned knowledge, has to be stored in more or less rigid structures.

But one must not forget that structure is adaptedness, not adaptation, knowledge already possessed, not cognition, not acquiring of knowlege. 'The word dies in the pen', wrote Goethe. 'An idea flows like molten lava', said Nietzsche, 'but lava builds up a fortress round itself, and all ideas ultimately suffocate themselves in laws.' As bone cannot grow without the dismantling of bone structure so human knowledge cannot advance unless what has already been adapted and is already known gives way step by step to be replaced by new and higher knowledge. And as genetic constancy and variability have to strike a balance in the genome of an animal or plant, so also do the invariance and the adaptability of knowledge in a particular culture. This chapter deals with the factors tending to preserve invariance.

10.2 *Habit and superstitious fear*

I have described in detail in my book *On Aggression* the part played by simple habit in the fixation of learned behaviour patterns. Individually acquired habits — a path habit, for instance — often assume a fixed form in a remarkably short time, and only with difficulty can the animal deviate from it, if at all. For the animal that possesses any causal insight into the possible consequences of its actions it is good strategy to keep strictly to a pattern that has proved successful and safe. I need only mention in passing the story of my 'superstitious' goose Martina — some would call her 'neurotic' — who first forgot her customary detour in her haste, then became afraid and went back to make the now superfluous detour as usual; or the story of Margaret Altmann's horse and mule which refused to pass a spot where they had camped several times in the past.

In human beings, too, acquired habits quickly become pleasantly comfortable, so that we find a departure from an established pattern unpleasant, even upsetting. I observed this for myself when I once tried to deviate from certain path habits that I had involuntarily adopted; I definitely experienced a slight anxiety. The typical neurotic compulsion which drives a man to go through strange, often highly contorted patterns of behaviour is only a hypertrophy of a mechanism whose normal, indispensable function is to preserve invariance, and which is indispensable to the storing of inherited knowledge.

The deep sense of anxiety that comes over an animal whenever it deviates from its habitual patterns of behaviour is a primitive impulse which was already powerful in pre-human times but is also essential to the complex motivation patterns of human civilization. The feeling is basic, for example, to feelings of guilt and thus encourages obedience to the law among civilized peoples. As I showed in *On Aggression*, there would be no possibility of reliable communication, no honest agreements, no loyalty and no laws if habit had not become a powerful motive in human behaviour.

However, no one would think of talking about 'desirable' habits and customs if, besides the fear of breaking them, there were not also other emotions involved, which reward us for faithfully carrying out established usage. We all know the peculiar sensation of pleasure we get from seeing something familiar again after a long time — a view we knew as children, the inside of a house we once lived in, the face of an old friend. A similar sense of satisfaction comes from the successful performance of a learned skill. This feeling, which derives both from the receptor and the motor side, is the very opposite of the feeling of fear described above: it is a comforting sensation of security, going well beyond the simple elimination of fear, for it greatly enhances our sense of self-confidence. So we say to ourselves, 'Obviously I'm on familiar ground here', or: 'I can still do that well'. We all underestimate, I think, the extent to which we are constantly haunted by anxiety and equally long for security.

10.3 *Imitation and tradition*

In civilized man all those processes, not specifically human in themselves, which help to consolidate what has become habitual are greatly intensified. Whatever we acquired from our parents and older relatives — and it cannot be said too often that we are civilized creatures by nature — we are bound to vest it with the emotional values that these older generations hold for us. If these values disappear altogether, cultural tradition will be interrupted.

It is not easy to analyse the nature of the various emotions which a young man has to feel for a member of an older generation if he is to be capable of receiving tradition from him. Moreover, the various qualities of these emotions can only be studied by phenomenological methods, so that, strictly speaking, a person can only speak about his own emotions. Contrary to my usual respect for the psychological subtlety of everyday language, I doubt whether we have an appropriate word for every single one of these emotions. How many different meanings there are to the word 'love'! Yet some kind of love must necessarily be present if there is to be tradition. Perhaps one needs at least to 'like' somebody if one is to receive tradition from him, for it is not easy to accept anything from a person one does not like.

The second of these emotions is sometimes called 'fear', but this too is a far from unambiguous concept. The use of the word in the phrase 'the fear of God' probably comes closest to the meaning we require. For as long as a young man's 'socialization' — that is, his integration into, and identification with, his cultural tradition — has not yet reached the point at which he senses in his blood that the richness of this tradition is something before which he should tremble, something to be revered for its own sake, it seems that he must needs adopt an older contemporary from within that culture as its representative and as the object of his personal aspiration. The notion of trembling reminds one of the Quakers, one of the finest and most rational of all religious sects. People today are not given to trembling or 'quaking' before a father figure, but anyone who inherits tradition must necessarily set the man from whom he receives it higher than himself. Perhaps the most appropriate word for this feeling of obligation and indebtedness is 'respect'.

It is a widespread error to believe that love and respect are irreconcilable. Thinking back to my childhood, I have tried to recall which children of my own age group and which of my older relatives and teachers I have loved most. Among the former I find that there were at least as many for whom I had felt respect, even a certain fear, as there were who had been faithful friends but also somehow my inferiors. I clearly remember that I have rarely so admired and been so fond of a friend as of the boy, four years older than I, who was our undisputed leader in Altenberg. In our early years at the Gymnasium we still used to play Red Indians with great enthusiasm, and I was in considerable fear of him — not surprisingly, since, as our leader, he punished all breaches of discipline, especially offences against the Indian code of honour. He was a scrupulously fair ruler, highly responsible and extremely courageous. He once risked his life to save the girl who is now my wife. To this man, Emmanuel La Roche, my first real master, I owe many of my ethical principles. And when I think back, I find that, even in those contemporaries of mine whom by the criteria of animal sociology I would regard as my inferiors, there was, and still is, something impressive, something which made me feel inferior to them. I doubt if it is possible to feel real affection for anybody who is in every respect one's inferior.

The correlation between affection and respect is even closer in the attitude of children to adults, while in that of adolescent boys to grown men it is virtually absolute. Almost without exception it was the strictest teachers for whom I had the greatest affection — 'strict', of course, signifying not some form of arbitrary tyranny but the legitimate demand that one acknowledge their superior position. As an exception, however, I must confess that I also remember two elderly maiden aunts who used to spoil us children shamelessly; they did not evoke any sort of respect in me, but I did feel a tender affection for them, mingled, perhaps, with a little pity.

Even the most elementary way in which tradition is preserved, namely imitation, assumes that the imitator is in some way impressed by the person he imitates. Sometimes this may go no further than the way my little grandson was struck by the Japanese ritual of bowing (see p.153). On a higher plane children try to project their whole personality into that of the person they are pretending to be — what we call play-acting or 'pretending'. The role a child chooses to act depends on what impresses him, and the pleasure he feels derives from the lift that his acting gives to his self-esteem, as I myself can well remember. Readers will not be surprised to learn that the parts I wanted to act were chiefly those of animals, and I felt very proud of myself pretending to be a duck or a wild goose. It also gave an ecstatic boost to my ego to pretend I was an express train locomotive, blowing its whistle and roaring through the countryside. These childhood memories make me think that the people or things we imitate between the ages of, say, eight and ten leave a deeper mark on our minds than those we imitate later, when we are not so completely immersed in the role we are playing.

For a boy to pretend he is an inanimate object like a steam-engine shows how wide is the range of his inborn capacity for imitation. In my own life 'pretending' has undoubtedly played a decisive role in one particular respect. I can still remember clearly how hard I tried, like a real actor, to imitate the actions of my favourite animals, thus achieving a high degree of empathy and understanding. A lasting result of this is my habit of trying to memorize the behaviour patterns of animals by imitating them. My talent for these histrionics amuses my pupils.

Children less pampered than I choose more obvious models to copy: bus drivers and conductors are still familiar figures, whereas soldiers, who until a few generations ago were the most popular of all, have since lost their power to impress. Koenig, Eibl-Eibesfeldt and others have shown that in 'lower', less complex cultures, the parts children most like to play are those of grown-ups engaged in activities that particularly excite the children, and, according to Koenig, they often pass naturally from pretence to actually helping grown-ups in the activity in question.

Elements of this childish love of 'make-believe' must also be involved when, as adults, we unconsciously ape people we acknowledge as superior to ourselves and whom we set out to emulate. In such cases we may copy, for instance, the way a man clears his throat or blows his nose. My wife and a number of observant friends have often told me that, when I come to a particularly important point in a lecture, I fall into the habit of speaking in the kind of metrical, somewhat staccato style of my teacher Ferdinand Hochstetter. I refused to believe this until I once found myself at an important lecture given by my pupil Eibl-Eibesfeldt. For, as his emotion grew, I heard to my astonishment echoes of the way Hochstetter used to talk — an acquired characteristic which had now passed to the second generation.

I need hardly say that Hochstetter influenced me in more ways than in mannerisms of speech. It is as difficult to copy only the essentials in the behaviour of a man whom one reveres and strives to emulate as it is impossible to regard a man as a model in some respects while despising him in others. The powerful urge to emulate a teacher only takes effect if one is able to revere him in every respect, but particularly with regard to his ethical standards. What one inherits from such a model are mainly norms of social behaviour, of moral attitudes. The guilt one feels at any infraction of these norms is closely akin to the embarrassment one would feel if one were caught in the reprehensible act by the 'father figure' in question. His mild disapproval, even in matters of skill and not ethics, can have the effect of a punishment. The worst thing that Hochstetter ever said to me about a purely technical error was 'Well, that's nothing to be proud of.' I hardly dare to think what my reaction would have been, had my teacher made a similar remark, not about my skill as an anatomist, but about some more serious backsliding. Conversely, the equally mild and parsimonious praise uttered by such a man is the strongest reinforcement I can think of.

One is bound to regard everything one inherits from a revered teacher of this kind, particularly in the realm of social conduct, with the same esteem and respect one feels for the man himself. This is a powerful force for the maintenance of cultural invariance, and as this stability is often insufficient if not totally lacking at the present time, many responsible people tend to regard such influences as unconditionally beneficial. But they are only so as long as they are properly balanced against the forces whose function is the breakdown and modification of structures, so that the adaptedness of the system to a perpetually changing environment is guaranteed.

All structures handed down by tradition display the stability essential to their supporting function. Every father figure has a father figure of his own, and this grandfather figure, whom the younger generation may not have known personally, appears as an even greater figure of reverence. As a result we have a phylogenetic programme of ancestor worship. Small wonder that ancestor cults are found in almost identical form among very different peoples. As the feeling of reverence grows with the passage of time, often to the point of apotheosis, so respect for traditional patterns of behaviour increase with their antiquity. Behaviour whose roots lie in the mysterious depths of the past assumes the quality of the divine, to offend against which is a sin, accompanied by feelings of fear and guilt.

Complementary to these processes, which involve punishment for offences against traditional norms of conduct, are those that reward obedience to established rites and customs. The possibility of identifying with a father figure and realizing that one is obeying the moral commandments of a super-ego, gives us an inner security that we cannot do without. One of the principal techniques employed in the diabolical practice of brain-washing is to destroy this sense of security by making the victim doubt everything that he thought he was sure of.

All these various forces move slowly but inexorably towards the point at which the body of knowledge proper to a culture and shared by all its members becomes consolidated in the form of dogma. Within certain limits, this process is absolutely indispensable. It is so even in that human activity which one would least expect to accept dogmas or to base its procedures on the firm rock of faith — namely science. Scientists may repeatedly tell themselves that everything we think we know is merely a working hypothesis, and that they are always prepared to admit, with no trace of emotional protest — indeed, even with delight — that everything they had previously believed to be true is false. This may well be so in the case of recent hypotheses that are the focus of present-day research. There may also be scientists who follow in the footsteps of Karl Popper by aiming at nothing but the invalidation of their own hypotheses so as to arrive eventually — by exclusion of all other possibilities — at the one unfalsifiable theory.

But, so far as I can see, scientists endowed with strong powers of Gestalt perception hardly ever proceed in this way. Their initial hypothesis is not an arbitrary construct arrived at without reference to external observations but invariably the product of the complex function of sense organs and central nervous system described in Chapter 7.2. When I observe my own procedure, I must admit that initially I believe what I have intuitively perceived. I then certainly do my best to disprove this belief by all possible means, giving it the subtlest chances to prove its falsehood or its truth. At this stage I can feel genuine pleasure if it turns out that my belief was wrong. But it would be untrue to say that I *want* all my hypotheses to turn out wrong, and in particular I would hope that my older hypotheses were capable of resisting all attempts to disprove them. I do hope, however, to detect minor inaccuracies in them, since I know that they cannot be one-hundred-per-cent correct — I am too well aware of the pitfalls of Gestalt perception to believe that. But I also know from experience that these powers very rarely tell me anything *completely* wrong, and I always count on their being at least partly right. By 'right' I mean what Father Adalbert Martini once gave as a definition of truth: 'Truth is that error which best prepares the way to the next smallest error.'

Without this particular way of hoping to be not too far off the beam, a scientist would hardly ever have the courage to take his new hypothesis as a foundation for further research. To 'suppose' something means, in its original sense, to lay down a new foundation. This procedure would be devoid of sense if one were not to expect the 'supposition' to be firm enough to carry a superstructure, and the greater the effort we put into erecting this, the greater the confidence we must have in its foundations. But, conversely, the greater also is the courage, and especially the industry, required to pull the entire structure down and rebuild it. Yet it is a decision that every scientist must be prepared to make.

10.4 *The search for identity*

Man, by nature a civilized being, cannot live without the supporting structure which his personality derives from the culture to which he belongs. The superficial imitative games of one's childhood gradually develop into spiritual emulation of a figure on whom to model oneself, and, having identified with this figure, one feels oneself to be part of his culture, even heir to it. Without this experience man can have no true awareness of his identity. Every old-fashioned farmer knows who he is — and is proud of it! The desperate search for an identity — a subject often discussed in our newspapers and a problem that besets modern youth — is a symptom of a hiatus in the continuum of our cultural tradition, and it is extremely difficult to help those who are caught up in this predicament. If a man loses contact with the culture in which he has grown up, and finds no intellectual substitute for it in another culture, he has no chance to identify with anyone or anything: he is a nobody, a nothing, as one can see in the despairing emptiness of so many young people's faces today. A man who has lost his cultural inheritance is indeed disinherited. Small wonder that he frantically takes refuge behind a protective wall of dogged autism and becomes an enemy of society.

No man can preserve his spiritual well-being without identifying with other men. But nor can he do so — as brain-washers know all too well — without receiving a modicum of appreciation and approval from others, and even those who are of sound mind sometimes anxiously ask themselves: 'Who am I?'

Because it strikes me as an instructive way of trying to understand this search for identity, I shall attempt to describe as objectively as possible what I experience in such crises of self-confidence. My first thought is certainly not to seek solace in my scientific achievements. Not that, even in such moments of depression, I doubt that my results are approximately correct — it is just that they seem unutterably banal. No — what saves me is the knowledge that by and large I am a man not unlike Ferdinand Hochstetter, Oskar Heinroth, Max Hartmann and others like them. There is a subconscious element in this which is akin to the way in which children pretend to be somebody else. It is as though I were playing at being Hochstetter, as I used to play at being a train or a wild goose when I was a small boy, and it gives a similar kind of lift to my self-esteem. I then recall that other men, greater than I, acknowledge the value of my work and treat me as their equal. These thoughts and feelings all come to me quite unsought and spontaneously — presumably they represent an innate programme in the species *Homo sapiens.*

There is a further and rather easy way, as I discovered, in which to restore my self-esteem. If my depression is connected with my work, making it appear uninteresting and not worth publishing — which regularly happens whenever I am about to finish a substantial manuscript — I read a book by one of my bitterest opponents. The more confused and violent it is, the more easily I can convince myself that there must be something in my own publications after all.

It is, of course, possible that most men are not quite as dependent on identifying with and being approved by others, as I am, and not as sensitive to their opinions. Yet even the greatest among us cannot stand entirely alone, nor is it desirable that they should be able to do so. Yerkes said that one chimpanzee is as bad as no chimpanzees. The same truth underlies Arnold Gehlen's remark that one man is the same as no men at all. The human mind is a supra-individual phenomenon.

10.5 *Phylogenetic ritualization*

There is a large complex of behaviour patterns, very diverse in origin but remarkably similar in function, which plays an important part in preserving the invariance of cultural tradition. This complex was discussed in part in the preceding section, but since it also has other functions, and since it is so vital to both animal and human behaviour, it needs a section to itself. The subject is what Julian Huxley called 'ritualization'. There are remarkably extensive parallels between these processes in the phylogenetic and the cultural field, so let us first turn our attention to the former, since this is the easier of the two of which to construct a functional model.

Over fifty years ago Huxley made the significant discovery that communication between animals of the same species, or, in objective terms, the coordination of social behaviour, is effected by means of signals which symbolize a particular behaviour pattern. In his classic study 'The Courtship Habits of the Great Crested Grebe', published in 1914, he described how the male grebe courts the female by fetching nest-building material from under the water, returning to the surface with the material in its beak and executing movements that are clearly identifiable as those of building a nest. It is as if it were saying to the female: 'Let's build a nest together'.

Huxley had already realized at that time that communication between human beings also often develops from the symbolic performance of certain actions. Since this development is not phylogenetic but cultural, and based on conceptual thought, human beings frequently create true object-independent symbols. Nevertheless the functional analogies are sufficiently close to entitle us to refer to ritualization and ritual acts, as Huxley did as early as 1914. My book *On Aggression* discusses the question of ritualization in detail, but I cannot simply refer the reader to the relevant chapters there because there are other important aspects of the subject that concern us here — in particular the functional analogies that make the two types of ritualization so strikingly similar.

I begin with phyletic ritualization, because this is a field where we know more about evolution than in most others, and also because it offers a simple model of ritualization in general. Our comparatively detailed knowledge of phylogenetic ritualization stems from the 'heroic' age of ethology in the early decades of this century. Ritualized behaviour patterns are a particularly good subject for comparative phylogenetic research, and it was here that Charles O. Whitman and Oskar Heinroth first discovered the possibility of a truly comparative science of ethology. Expressive or display movements, ritualized means of communication, together with the gradations of similarity and dissimilarity in these actions observable in various species, breeds, phyla and classes — this, in the first instance, was what gave Whitman and Heinroth the idea that behaviour patterns could be as reliable indications of generic relatedness as physical characteristics, and this discovery meant the birth of ethology proper.

Thus ethology came into existence as a subsidiary discipline that provided general phylogenetic research with valuable taxonomic data. It received in return important information concerning the phylogenetic evolution of inherited motor patterns, and also, because expressive gestures and actions are for many reasons particularly profitable subjects for comparative study, valuable fresh knowledge about the emergence of behaviour patterns of a quasi-symbolic character. In a considerable number of animal groups we know there are differential sequences of homologous behaviour patterns whose development we can trace. They extend from the unritualized prototype, i.e. a motor pattern not yet adapted to communication purposes, through numerous intermediate stages to highly ritualized behaviour which has been modified out of all recognition by the selection pressure of its communicational function, and whose origin one would not suspect if fortune had not presented us with the complete sequence of intermediate steps.

Phylogenetic and cultural rituals have four essential common functions, and it is these which give the results of ritualization the unmistakable hallmark of formal analogy:

1 The first and oldest function is that of communication.
2 The second, which in the case of phylogenetic ritualization probably developed from the first, consists in the 'channelling' of certain behaviour patterns into specific areas as a result of their ritualization, in the same way one can channel a river in the direction one requires. In phylogenetic ritualization it is principally aggressive behaviour that is channelled in this manner; in the cultural process it is virtually the whole of social conduct of both phyletic and cultural ritualization.
3 The third basic function of both phyletic and cultural ritualization is the creation of new motivations which actively influence the complex of social conduct.
4 The fourth function is the prevention of interbreeding between species or pseudo-species, i.e. between cultures and subcultures.

A further function, proper only to cultural ritualization, is the creation of independent symbols which become established in society and are defended as part of that society.

Let us now compare these four functions.

(1) We first turn to the communicative function of phyletic ritualization. Every system of communication consists of two complementary parts — a transmitter, or sender, and a receiver. In other words, a receptor mechanism has to respond selectively to a signal, a key stimulus. In phylogenetic ritualization development proceeded from the receptor end — i.e. survival-purposive responses of some survival value evolved which were elicited by actions that a fellow member of the species would have performed in any case. This phenomenon has been known for a long while under names such as 'resonance' and 'social induction', but the question has never been asked: What are the physiological mechanisms that make a horse panic when it sees another horse charging past, or that make a hen which is almost sated start to eat again when it sees a hungry fellow bird pecking?

It is the development of a receiving apparatus that makes it possible for one animal to 'understand' the behaviour of another, and turns an action into a signal. It has been termed 'semantization' by W. Wickler. What has previously had only one specific survival value, like escape responses or eating, now acquires a new communication function and is understood by its fellow animal. But initially semantization does not change anything in the action pattern itself. Where ritualized behaviour emerges, however, it is probably the first indispensable step in the evolution of any phylogenetically ritualized behaviour.

This new communicative function, or, to be precise, the new receiving apparatus brings its own selection pressure to bear on the 'transmitter'. All the characteristics of the behaviour pattern that serve to strengthen its signal function are very valuable from a selection point of view and therefore become overstressed as evolution proceeds. In addition, the signal thus evolving frequently becomes visually emphasized by various physical structures. The new selection pressure, of course, is in opposition to the original survival function of the behaviour pattern, which is threatened by any sort of change. Only in the case of non-functional epiphenomena like displacement activities, or 'gestures of embarrassment', intention movements and autonomic activities, does this opposition not apply, which is exactly why it is from these that the vast majority of expressive or emotional movements have evolved.

Only rarely do signal actions come to be grafted on to functional behaviour patterns, and even more rarely do such changes by ritualization affect the fixed motor pattern itself. One of these exceptions is the wing beat of pigeons, the amplitude of which increases in courtship and in taking off, to such an extent that the primaries come into contact both above and below, producing a loud flapping noise that acts as an acoustic signal. As a rule, however, the new communication function cannot usually be combined, as in the pigeon, with the original unritualized function, and the signal becomes an autonomous motor pattern in its own right.

(2) The second function of ritualized behaviour is to direct the behaviour patterns of a particular species into specific channels and especially to prevent, or at least minimize, the adverse effects of aggressive behaviour within that species. We have seen above that changes caused by the development of the communication function may weaken the original effectiveness of the action and thus impair the chances of survival. Intraspecific aggressive behaviour is an exception, since here it is altogether desirable that the effectiveness of the original motor patterns should be considerably reduced and with it the danger of physical injury to the combatants. In the vast majority of animals intraspecific aggressive behaviour has derived from feeding behaviour, and most fishes, reptiles, birds and mammals use their feeding organs, mouth, teeth and (in carnivores) paws for fighting; only relatively few plant-eaters use their front feet, and fewer still — reptiles, for instance — use their tail. Even rarer are the cases where organs and behaviour patterns evolved as a defence against predators are used in intraspecific aggression. In fact the only cases that occur to me are butterfly fish (*Chaetodontidae*) — which use the spines on their dorsal fin — and certain horned ruminants. The only animals whose weapons have been developed solely for rival fighting appear to be stags.

Since the purpose of intraspecific fights, as far as the survival of the species is concerned, is not to kill but to subjugate the rival or drive him out of the territory, the weapons and behaviour patterns for killing a prey or for defence against predators are far too powerful, cruel and effective for rival fighting and there is thus a high survival value in limiting and channelling their effects. This is achieved in many animals through displays of antagonistic behaviour which precede the actual fighting. In most cases this behaviour has evolved from intention movements and ambivalent behaviour patterns derived from instinctive conflicts. Such behaviour often becomes highly ritualized. Ritualized action patterns of threatening and intimidation often evolve and we find two rivals literally measuring up to each other: in the so-called broadside-on display, rival fish measure the size of their bodies against each other, and when they fight with their mouths, they test each other's strength. A 'matching' of strength also occurs in fights between stags.

As already stated, all signals evolve under the selection pressure brought to bear upon them by the corresponding receptor mechanisms. Mimicry, i.e. the imitation of a signal by another species, consists of the unilateral adaptation by the sender of the signal to the receptor mechanism of the species being imitated, and this, as Wickler has shown, makes it a particularly simple example of how signals evolve. When the sender and the receiver belong to the same species, as is usually the case, they are both equally subject to the selection pressure of their communication function, complementing each other even more clearly and strikingly.

The appearance and differentiation of an independent ritualized behaviour pattern come about by its being split off from its 'unritualized' model, which retains its original non-communicative function. Therefore, every higher differentiation of a communicative action implies the coming into existence of an altogether new, autonomous fixed motor pattern, which, like every other, possesses its own spontaneity and its own appetitive behaviour. In other words, the animal *wants* to perform this sequence of movements; the performance has become a *need*. In the greylag goose, for instance, the so-called triumph ceremony has become the central bond which holds individuals together. Mates, parents and children, as well as siblings, participate in it, and an individual deprived of this social link falls into deep depression. The triumph ceremony not only symbolizes the union of individuals: it actually brings it about.

As W. Wickler and his co-workers have shown, the bond by which two or more individuals are kept together consists — for mammals, birds, fish and even crustaceans — in particular ritualized activities which one individual can only perform with one definite individual partner. Only in the case of the monogamous shrimp (*Hymenocera*) is this not so: the male and female are here united by a common urge to attain a state of quiescence, and when the male discovers t' he has lost his mate, he searches everywhere until he has found her again, whereupon they both subside into a state of utter tranquillity.

One of the most remarkable examples of a ritual of this kind is the duet sung by gibbons, barbets, shrikes and drongos. The two partners sing short snatches alternately in rapid succession to produce a continuous melodic line which no one listening would suspect of being the work of two different individuals. The fragments and the way they are linked together differ considerably from pair to pair, and it is as though the partners must have learned to attune to each other — rehearsed, as it were — in order to achieve this harmonious result. The subject needs to be further investigated, but if this is so, then each bird would only be able to sing this duet with one other member of its species, and the urge to perform the ritual would create a powerful bond between the partners.

(3) The process of phylogenetic ritualization thus gives rise to a new and autonomous motivation participating in the manifold interactions of social behaviour. The new ritualized behaviour pattern attains a seat in what I called, in *On Aggression*, 'the great parliament of instincts'. With many social animals the structure of the community is largely determined by ritual behaviour. The triumph call of the greylag dominates the entire social life of the species. With gannets the whole arrangement of the nesting colony, including the precise distance between the nests, depends on rituals so highly stylized that it is difficult to ascertain their phylogenetic origin. Similar stylized behaviour is found in jackdaws and many other social animals.

In higher social animals ritual behaviour patterns, with their dual function of communication and the motivation of social behaviour, form a unitary integrated system which, though highly regulative and adaptable, is firm enough to determine the whole social structure of the species in question. Frequently both the rigidity and the regulative force of such a system derives from the tension between opposing rituals, such as those of threatening and those of appeasement. One need only observe baboons in the zoo for a little while in order to see how these two functions balance each other. Similarly, in a community of wolves or chimpanzees the great majority of expressive actions and gestures reflect an urge to threaten or appease, and it can hardly be an accident that in normal circumstances one very rarely sees open violence among members of such aggressive species.

(4) A fourth essential function of phylogenetic ritualization is mentioned here as an appendix, as it were, and only because it also has its analogy in the sphere of cultural development. Ritual behaviour patterns may contribute to prevent interbreeding. Many courtship patterns do just this. In certain manakins (*Pipridae*), for instance, small tropical birds, the male is brightly coloured but the female differs little from one species to another. Chapin and Chapman have proved that the males are also attracted by females of other species but that the latter respond with strong selectivity exclusively to the courtship of males of their own species. As Heinroth pointed out many years ago, the same is true of the social courtship of many species of dabbling ducks.

10.6 *Ritualization in culture*

In the field of human culture, as in that of phylogeny, ritualization can be regarded basically as the development of a system of communication in which the formation of symbols represents a decisive step. The phylogenetically evolved signals discussed in the preceding section cannot, however, be equated with symbols in the cultural sense, for the former, acting as 'releasers', cannot be employed at will, nor can their significance be learned, since the whole apparatus of communication is phylogenetic in origin and controlled by the processes of heredity down to the last detail. Learning plays only a small part in the social activities of animal groups and has not the slightest influence on the form of the mechanisms that send and receive messages.

Yet despite these differences the phylogenetically evolved signal and the culturally evolved symbol do have one thing in common: they both originate in the emergence of a creature's ability to 'understand' those behaviour patterns which allow it to predict how a fellow creature is going to react. Typical of such patterns are so-called intention movements — that is, incomplete movements which indicate that an animal is preparing to perform a particular action. In phylogenetic ritualization this 'understanding' derives from inherited functions in the receiver, and the behaviour patterns 'understood' in this way are fixed motor patterns. In the cultural context, on the other hand, sending and receiving signals both depend on learning and the inheritance of acquired characteristics.

So far as the sender of the signal is concerned, the ability to imitate one's own behaviour — a faculty nascent in anthropoids but fully developed only in humans — enables the individual to exhibit a copy of the particular pattern that the sender wishes to convey to the receiver. As already discussed (Chapter 7.7), this imitation presupposes the existence of freely available voluntary movements. Cases have been recorded where a chimpanzee has used imitations of particular intention movements to induce its partner to join in a certain activity. In the Yerkes Laboratory two chimpanzees were given the task of pulling up a basket by means of a piece of string threaded loosely through the handle. The two animals had to pull the ends of the string at the same time. When one of the chimpanzees saw how to solve the problem, he took his companion to one end of the string and made him hold it in his hand; he then ran quickly to the other end, picked it up and mimed the action of pulling it. This seems to have been the nearest that an animal has come spontaneously, i.e. without deliberate conditioning, to the use of a true symbol.

In his interesting speculations on the origins of spoken language, Gerhard Höpp came to the conclusion that the first real utterance must have been an imperative, and the few observations that have been made of the ways in which elementary symbols have come about seem to support him. Among the higher animals the commonest situation in which the need to communicate with others arises is when one requires help. For a thirsty dog to nudge its master with its nose, stand on its hind legs against the wash basin and look back at him over its shoulder, is something that can happen only under the pressure of great urgency, and only once have I seen the cleverest of my dogs behave in this way.

It is significant that the most elementary acts of human communication which we can take for certain as not being innate are of a similar kind. Helen Keller, blind, deaf and dumb, who had no intellectual contact with human beings until the age of seven, was able, even before Anne Sullivan began to teach her, to make clear her need for food and drink by imitating the relevant actions. In this case the imitation can only have been self-imitation.

As a culture develops, such first beginnings of communication become differentiated in a manner similar to that of innate signal mechanisms. The other three survival functions of phylogenetic ritual that serve to perpetuate the species — the direction of behaviour into non-destructive channels, the formation of new motivational patterns, and the prevention of crossbreeding — also have parallels in human culture. We may note here and now, however, that whereas phylogenetic ritualization has no influence on the invariance of the characteristics of a species (except indirectly by preventing crossbreeding), cultural rituals play an important part in preserving the traditional characteristics of a culture.

In the development of systems of cultural communication the characteristics of the sender are in the first instance determined by the requirements of the receiver. Hence we find in cultural rituals virtually all the characteristics we have seen in the phylogenetic context, to ensure clarity and precision. For a signal to have a precise meaning the receptor mechanism must necessarily possess a certain selectivity, and this is far smaller in innate releasing mechanisms than in learned responses. The ability to distinguish one complex of stimuli from another, even if the difference lies not in the composition but only in the configuration of the components, depends on perceptual functions that belong to a far higher level of the central nervous system than do innate releasing mechanisms. Furthermore, learning processes also have an important part to play here.

Although in all cultural communication systems it is learned recognition of Gestalt that plays the role of the receiver, simpler processes of perception, taking place on much lower levels of the nervous system, are also involved, as they are the elements and the base of more highly differentiated Gestalt perception. Physiologists and psychologists who have concerned themselves with these functions know very well what kind of demands our perceiving apparatus puts to the constellations of sensory data, if it is to recognize them as unmistakable Gestalts. Unambiguity is regularly achieved by the union of greatest possible simplicity and greatest possible general improbability. On a lower level, the same demands on innate releasing mechanisms are being fulfilled by 'releasers', e.g. by the equally innate stimulus-sending contrivances. In spite of the immense differences in complication, both types of receiving sets contain the same elementary functions of perception — which explains their similarity. For the same reasons, as I have shown in my study 'Stammes- und kulturgeschichtliche Ritenbildung', similar types of receiver have caused parallel characteristics to develop in the senders from which they receive their signals.

Let us now turn our attention to the four parallel functions of ritualization in the field of human culture.

(1) There is little that need be said about the communicative

function. Virtually all means of verbal communication derive from ritualizations. Even displays of emotion, which contain a large proportion of innate components in all cultures, have traditional ritualized elements superimposed upon them. As with phylogenetic behaviour, the original function of cultural rituals was probably that of communication, and the other functions can be derived from this.

(2) While in the phylogenetic context the second function, that of controlling and checking potentially dangerous behaviour, is basically restricted to removing the lethal aspects of fighting behaviour; in human affairs ritualization affects virtually all social conduct. Practically all behaviour patterns which are really untouched by any sort of cultural ritualization — e.g. instinctive actions like scratching oneself, picking one's nose, stretching, or similar 'comfort' activities — are socially taboo, in the same way as are defaecation and copulation. One of the immediate products of this all-embracing ritualization is the emotion of shame.

What function furthering the survival of human culture and with it of the human species is served by this strait-jacket of ritualization into which most of our social intercourse is so uncompromisingly pressed? It probably rests on the necessity of bringing most or all of the instinctive urges inherent in our species under sufficient control to impose upon them the normative pattern drawn up by the culture in question.

As cultural pseudo-speciation moves very much faster than biological evolution, the older and more advanced a culture becomes, the greater the discrepancy grows between the innate norms of social conduct that a person inherits and what his culture requires of him. As I have said, this may be one of the reasons why cultures regularly collapse when they have reached the stage of 'high cultures'. This process, which Spengler equated with dying of old age, evidently does not affect simpler cultures which remain at a lower level of development, 'closer to nature', like the Pueblo Indians of New Mexico, whose traditions stretch back into prehistoric times.

Whereas phylogenetic ritualization is restricted to social behaviour among conspecifics, cultural ritualization also influences man's behaviour in his dealings with his extraspecific environment. Within this environment man surrounds himself with a self-made world of objects which encloses him like a shell, and sometimes prevents him from seeing that outside this man-made shell there is an objective reality independent of man. Many writers, among them Arnold Gehlen, have made the utterly misleading claim that man has no such environment.

In Hans Freyer's book *Schwelle der Zeiten* there is a chapter called 'The Triumph of the Object', in which the author analyses with remarkable perspicuity the part played in the life of society by the concrete, man-made object, the artefact — this being the sense in which Freyer uses the word 'object' (a far narrower sense, incidentally, than that in which I have used it in Chapter 7). 'While all activities repeat themselves in an endless cycle of needs and their fulfilment,' writes Freyer, 'the creation of objects has a definite beginning and a definite end.' Such objects are not consumed by constant use; they are not imperishable but durable, 'surviving in the sense that the processes of decay and disintegration to which they are subject affect only their material aspect, not their essence'. These words of Freyer's define the immortal nature of the object made by mortal hands, an immortality akin to that of the Platonic ideal. This transcendental pattern of human creativity is probably the paradigm for all idealist conceptions of the process of creation.

The triumph of the transcending artefact culminates in releasing the object from its original utilitarian purposes and conveys to it a right to exist which is based exclusively on spiritual values which, as Freyer says, 'are instilled into it in the act of its creation, and which, in it, have become an objective phenomenon, existing as such. This is the case with the work of art'. Further on: 'Only when the human mind turns to the creation of artefacts is the category of the aesthetic constituted.'

I have my doubts about this, since I suspect that the aesthetic dimension, the new and constitutive property of art, emerges far earlier in the history of human behaviour, namely, in what is undoubtedly the earliest form of art for its own sake: the dance.

Be that as it may, the world of man-made artefacts, man's clothing, furniture, houses and gardens, the culturally changed landscape of orderly fields, woods and vineyards — the magic landscape, as Freyer calls it — and, most of all, his works of art, which surround man on every side, inevitably leave their mark on cultured man and find their expression in his behaviour. From the superficial conventions of manners to the underlying substance of ethical attitudes and convictions, social conduct bears the mark of the age, and the spirit of that age imposes on man's innate programme of social conduct a pressure that increases with the development of the culture in which he lives. One of the reasons why high cultures suddenly collapse may be that a revolt breaks out against a situation in which a culture that is becoming more and more ritualized imposes a degree of constraint which is felt to be increasingly intolerable — a revolt diagnosed as a 'decay of morals'.

It is a remarkable fact that preoccupation with external objects, together with manners and moral attitudes, can also influence man's physical appearance, his 'phenotype'. The concepts of structure and function cannot be kept strictly apart, and postures can become genetically fixed, thus forming part of the stock of visible characteristics of a particular breed. In the wild form of our domestic chicken, the tailfeathers are carried horizontally as in pheasants, and the rooster only raises them to an upright position as part of his demonstrative or 'display' behaviour. When he does so, he looks just as our domestic roosters do all the time, not because the latter are morphologically different but merely because they are over-sexed and perpetually in the mood for 'showing themselves off'. A certain physical posture prescribed by society can produce similar effects. The remarkable extent to which fashions in clothes influence the appearance of the body can be seen by comparing fashion advertisements with photographs from the same period. Even when fashion models are photographed in the nude, they take up poses that match the style of the clothes being worn at the time.

It is not just clothes but all the objects within a culture that bring an influence to bear on the behaviour and physical appearance of those who belong to that culture. As Hans Freyer has so convincingly shown, the noble knights and ladies of medieval romance could not have moved naturally in their high gothic halls, nor sat in their high-backed gothic chairs, if their habitual attitudes and postures had not been deeply influenced by the style of their surroundings. I use the word 'naturally' quite intentionally. Man is by nature a creature of his own culture and this implies the inborn readiness to comply with the demands of ritualized behaviour until they become what is rightly termed 'second nature'.

To wear the prescribed costumes of an age with grace and dignity is regarded as one's duty, though at times it must have been torment. In addition, the performance of ritualized behaviour patterns has been seen in most high cultures not only as a duty but also as a status symbol, and was therefore nurtured by innate sources of demonstrative behaviour and display. As we have seen, all learned behaviour patterns and acquired skills can become an end in themselves and a source of pleasure, but what human conduct ever attained that acme of sophistication and perfection represented by the patterns of social intercourse which in high culture have attained the character of true works of art? The *tenue* that a man observes and masters not only *seems* to be genuine and natural; it actually *is* to him the most natural thing in the world and in the interaction of his motivations it plays its part much as if it were phylogenetic in origin and genetically programmed.

In 1514 Count Baldassare Castiglione, professional courtier and authority on court etiquette, wrote a book on that subject, under the title *Il Cortegiano*, which I regret I know only through quotations by Hans Freyer. Castiglione was intrigued by the phenomenon that the courtly manners which one would suppose to be anything but a sincere expression of a man's feelings — and indeed rarely are — can become an essentially genuine expression of his personality. What seems, at its face value, only the most superficial veneer of a cultured exterior may really be a quality of true *humanitas*: outward decorum and inner rectitude may fuse into one indivisible unit, at least so far as the processes of their phyletic origin and their tradition are concerned. Both are based on an open phyletic programme which is then realized by the individual culture to which a man belongs and, with both, the innate base consists in the appreciation of ethical and aesthetic values, in other words, in functions of Gestalt perception.

One often hears hackneyed statements concerning the alleged inanity of good manners. As a rule, a person with ingrained good manners cannot be really rude, however hard he tries. I have watched some of these pathetic attempts. On the other hand, I know from bitter experience that it is a common trick of cruel and congenitally malicious people to camouflage their spiteful remarks as honest outspokenness, saying the most perfidiously heartless things all 'for the good' of the listener.

The narrow paths into which virtually all social conduct is channelled by cultural ritualization also serve to curb man's aggressive tendencies, though in general only when they are directed at members of his own culture and his own social class. This can lead to paradoxical and sometimes deplorable results. In the Middle Ages, for instance, the nobility of the various petty states formed one class, the peasants another. When one state waged war against another, the knights fought each other according to the rules of the joust, and scarcely more lost their lives on the field of battle than in these sporting contests. But the peasants were expected to fight the war for their respective lords in unritualized and bloody battles, and whereas conquered knights were received like guests, captured foot soldiers were treated like cattle.

In small cultural groups the direction of aggressive behaviour into particular channels is achieved by a remarkably wide range of rituals. According to Eibl-Eibesfeldt, parents among the Waika Indians openly encourage their children to fight, at the same time exhorting them to follow a strict ritual in the way they strike blows at each other. African bushmen, on the other hand, effectively educate their children to behave peacefully. The two peoples live in different ecological conditions, and whilst the former frequently wage war against neighbouring tribes, the latter hardly ever do.

In many cultures, including some of a relatively primitive kind, ritualization has converted aggressive tendencies into the form of contest that we call sport. This question is fully discussed in my book *On Aggression*.

(3) Little needs to be said about the third function of ritualization — that is, the creation of new and independent factors in social conduct. It is a matter of course for every member of a culture to take pride in the demonstration of skills and of 'good manners' (*tenue*) and to regard the objects of the man-made world around him, works of art and the ethical code of his society, as the highest values in that culture. The urge to defend and promote these values becomes a powerful motivating force dominating behaviour.

(4) The final function in which there are analogies between phylogenetic and cultural ritualization is that of ensuring the cohesion of the group and distinguishing it from others. As mentioned in the section on pseudo-speciation (Chapter 9.3), even the tiniest ethnic groups or subcultures are held together by a pattern of ritualistic norms but at the same time separated thereby from other comparable groups. Even in communities which are united, not by common cultural symbols but only by the bonds of personal friendship, like greylag geese or young children, the cohesion of the group is appreciably strengthened by its antagonism to other groups. In larger communities held together by common cultural symbols, the correlation between cohesion within the group and hostility to all others is still more pronounced.

Chapter eleven
Culture and change

11.1 *Man's persisting open-mindedness and curiosity*

As osteoclasts and osteoblasts work against each other in the creation of bone structure (see p.198), and as genetic change and invariability are in fruitful opposition in the evolution of species, so in the life of a culture the forces that make for the preservation of the structure of that culture are opposed by those, equally essential, that are concerned with its destruction.

How dependent each culture is for its survival on a balance between these two groups of factors emerges particularly clearly when we observe what happens if the one dominates the other. It can be as disastrous for a culture to become petrified in a set of ritualized procedures as for it to abandon all its traditional values and the knowledge that these values enshrine. So the trends of human behaviour to which I turn now are by no means 'bad' in themselves.

In the section on exploratory behaviour (Chapter 7.6), I pointed out that it is a characteristic of man that, unlike other higher organisms, he does not lose his urge to explore and play when he attains sexual maturity. This, in conjunction with his predilection for self-exploration, makes man constitutionally incapable of ever submitting entirely to the force of tradition, and in every one of us there is a tension between the pressures brought to bear by the hallowed values of tradition and the rebellious spirit of curiosity that makes us seek after novelty. *'Novarum rerum cupidus'*, was how the Romans defined the political revolutionary: 'A man eager for new things'.

As in the physiology of our endocrine glands, so also in the pattern of our behaviour every urge is combined with a counter-urge to produce an equilibrium — an 'equi-potential harmony', as some call it. Two opposing forces — on the one side adherence to tradition, with the accompanying feelings of guilt when one breaks with this tradition, and on the other side an equally strong and equally emotional desire for truth and new knowledge — can sometimes find themselves locked in fierce inward battles, battles fought at great cost to the man whose mind is the battleground. The greater the mind, the higher these costs seem to be, as we know of one of the greatest men, out of his own mouth. When Charles Darwin had opened the path to humanity's real understanding of organic creation, he did not feel like the great deliverer that he was; but wrote that he felt more like a murderer. On a less heroic plane the same kind of battle is waged repeatedly in the soul of every one of us. Individuals differ greatly, and the point of equilibrium between the conservative and the revolutionary tendencies varies from one person to another. Those with a great deal of confidence in their intellectual ability, who are often far from being the cleverest, frequently have a restricted emotional faculty and only a limited capacity for grasping a complex gestalt. Such people generally have no respect for tradition and are particularly prone to 'technomorphic thinking', a mental sickness I shall discuss in my second volume. Emotional people with a capacity for affection and respect often do not venture to embark on a rational criticism of traditional values, in spite of their excellent powers of analytical thought, and regard anyone who does so as a heretic bent on destroying the highest values of civilization. So, paradoxically, we may find highly intelligent people passionately and stubbornly opposing all causal explanations of natural events and all innovations in cultural and social life.

The culture-preserving balance between the factors that maintain tradition and those demolishing it may be in a state of happy equilibrium. The two pans of the scales may bear equal weights, but the volume they are asked to carry may differ greatly from one individual to another. In Darwin the tension between the opposing forces was obviously exceptionally strong, and it is quite possible that such strong tension favours productivity.

When one observes the interplay of the conservative and the progressive factors in one's own self one seems to find — as many who, like me, have reached their three score years and ten, will confirm — that one's conservative side is that of an old man, and one's progressive side that of a young man. In strict phenomenological terms I have to confess that I can still observe in myself, despite my advanced years, a puckish streak which is at odds with all professorial dignity and keeps urging me, particularly on formal academic occasions, to indulge in boyish pranks at the expense of custom and decorum. Nor am I the only one who feels like this. Once, when I was walking in full academic regalia in solemn procession with the members of the Bavarian Academy of Sciences, I suddenly got a well-aimed kick in the backside from a Nobel Prize winner who was walking behind me. A spirit of roguishness is, of course, quite without respect for tradition, including scientific tradition, and takes a slightly wicked pleasure in finding that something that has long been taken to be true turns out to be false, even if this makes for a great deal of further work.

At the same time there is a second spirit in me, one that right from my early days sincerely respected all forms of tradition, hung on the words of my teachers and is devotedly attached to all outward conventions and customs, including things like the wearing of academic robes. Although both spirits were present in me from the beginning, I am quite certain that the second has become stronger over the years. But at the same time I hope and believe that the bad boy within me will never completely die out.

11.2 *Youth and the desire for change*

In a beehive, different behaviour patterns are performed in the common interest of the stock by bees of different age groups. The young workers tend the brood, feed them with glandular secretion and produce wax; the elders fly out and forage for food. Potentially both age groups can perform both functions, for Rösch has shown that the young workers will forage for food if one takes away all the older bees, and vice versa that, if the young are not present, the old will not only tend the brood but even reactivate their atrophied salivary glands in order to feed the larvae.

Similarly, in human affairs there is a division of labour between different age groups. It appears so natural that older people should in general be conservative and younger people be looking for change, that we hardly pause to ask ourselves whether there might not be some underlying purpose and harmony behind this antagonism.

A spirit of protest among the young is far from restricted to human society. It is also found in hierarchically organized animal communities in which young and old live side by side for a considerable time. Young male wolves, for example, only begin to rebel against the leader of the pack when they are strong enough to take his place. This rebellion against a hitherto unchallenged leader frequently erupts with an unexpectedness that smacks of treachery, as men have found to their cost who have brought up a wolf, or an animal from a species with a similar social organization, in the environment of a human family.

In chimpanzees, as in apes in general, sexual maturity begins immediately after the animal has cut its second teeth, usually in its seventh year, i.e. before it attains its maximum weight. A further five or six years then elapse before the young male achieves the social standing of a fully mature adult. In man the process extends over a considerably longer time, and it seems reasonable to assume that it is the need to acquire a body of traditional knowledge that has produced the selection pressure to make this period so much longer. The two words 'childhood' and 'youth' have developed in naturally grown language to denote two qualitatively different phases of human growth. It is legitimate to speculate on the nature of their difference.

Man's extended childhood enables him to learn — that is, to fill his mind and his memory with the accumulated knowledge of tradition, including language. The equally extended period between puberty and adulthood — youth — also has its specific role. It seems a natural development, rooted in the phylogenetic pattern of human social behaviour, that young people who reach the age of puberty should challenge the traditional values they have inherited from their parents and look for new ideals. This is equally true of 'well-behaved' adolescents whose relationship to their parents does not show any outward change initially. Nevertheless, there must be a hidden waning of their feelings towards their parents and others who command their respect, and, as Bischof has shown, this extends beyond family and superiors to include everybody and everything familiar to them. What had formerly been so new and strange that they shied away from it, and were so afraid of that they even lost their curiosity, now becomes a sudden source of fascination. At the same time there arises a new boldness and aggressiveness, in the broadest sense, especially in the young male, and probably as a direct result of hormonal influence; this, combined with the urge for new knowledge and experience, leads in its turn to a spirit of adventure and *Wanderlust*. Wild geese exhibit the same pattern of behaviour. Bischof demonstrated that families of white-fronted geese broke up when the positive valencies attaching to certain members of the species suddenly became negative valencies. This process also prevents the mating of siblings.

In humans the changes caused by the onset of puberty are far more pronounced in the male than in the female, and a young man rebels more violently against his father than a girl against her father or mother.

Our faithful adherence to traditional norms of behaviour receives support from the fact that the feelings we have towards the person from whom we learn these norms also become transferred to everything that he teaches us. For most young people this person is their father, though in the large family units anthropologists have come to regard as primitive it may also be an elder brother, a cousin, an uncle or some other relative. Traditional values are, of course, also transmitted through the influence of groups of people *en bloc*, but I believe that in most cases there is one particular person who acts as a kind of tradition-bearing father figure.

Parents and old people who tend to reproach the young for their 'disloyalty' must realize that they would never be able to sever the bonds holding them all too strictly to tradition unless the reverence that the adolescent feels for the givers of old tradition suffered a slight diminution at a certain period of his adolescence, and, even further, an emotional change, turning attachment into aggression bordering on hostility, or, to be more precise, an ambivalent mixture of these negative emotions. The intensity of these changes depends on many circumstances, yet it seems certain that a dogmatic, hard-line educator will arouse more rebellious feeling than a mild 'democratic' one. However, the entire absence of all aggressive feeling makes the loosening of family ties, which is indispensable for normal cultural development, very difficult if not impossible.

Once an adolescent has begun to question and oppose the norms of social conduct exemplified by his father figure, he starts to look outside his family circle. This search may take the form of actual travelling, or of a purely spiritual journey of exploration. What sends young people out into the world is the longing to find something vague and splendid that is essentially different from the everyday events of family life. It is not too difficult to find a convincing answer to the question wherein the survival value of this search is to be found. It lies in the adolescent identifying with another social group whose cultural norms are sufficiently different from those of the parental tradition to effect something like a cross-fertilization, yet not so different as to destroy all respect for the former father figure and the values it represents. The search may often lead an adolescent to 'adopt' an older friend, a teacher or even a whole family, as his new contact with tradition.

In their critical stage of development adolescents regard their parents' norms of behaviour as dull, boring and out of date, and are ready to adopt others. An essential element in these is that they must embody ideals that one can fight for. This is why so many emotionally fully developed young people join minority groups which they regard as being unjustly treated and for whose rights it is worth fighting.

The astonishing speed with which they manage to attach themselves to a new cultural group, the fixation of their instincts for collective enthusiasm on a new object, have features which recall the biological phenomenon of imprinting (see p.78). Both phenomena are linked with a sensitive phase in adolescent development; both are independent of conditioning processes, and neither can be reversed, in the sense that a first union of this kind can never be followed by a second of equal intensity and power. Falling in love is another example of object fixation: the very phrase symbolizes the suddenness of the process.

When a young man finds in a teacher, an older friend or a group the embodiment of the new ideals for which he is longing, he may indeed develop feelings of rapturous worship whose symptoms are similar to those of being in love. However, it would be just as wrong to see this as a homosexual tendency as it would be to interpret his antagonism to his father as a symptom of an Oedipus complex. An entirely normal boy can experience this kind of rapturous reverence for a fat old man with a white beard. Every man has either experienced these things himself or observed them in others, and psychologists and psychoanalysts are familiar with the processes involved. But my own interpretation of them diverges radically from that of the analysts. My hypothesis is that in their controlled chronological sequence these processes are phylogenetically programmed, and that their survival value for the culture as well as for the species consists in removing out-of-date behavioural factors and replacing them with new, thereby ensuring a constant adaptation to the ever-changing conditions of the environment.

The higher a culture, the more essential are these functions for its survival, since as it develops, so the speed with which it transforms its own environment also increases. In general the replacement and modification of traditional norms appear to make a culture flexible enough to keep pace with environmental changes. There is good reason to believe that in old and primitive cultures tradition was more closely adhered to, and that sons followed more faithfully in their father's footsteps in such societies than in high cultures at their peak. Whether high cultures have actually collapsed because of the disequilibrium of the functions of preservation and demolition, more particularly because the forces of destruction have gained the upper hand, is hard to tell. Our own civilization, however, is clearly in danger of disintegrating as a result of an over-eagerness to change, even destroy, the whole of tradition.

In 'normal' circumstances and in a 'healthy' culture — terms that admittedly need definition, which is reserved for a second volume — there is a special in-built force that prevents this destruction of tradition and the loss of all the knowledge it embodies. When there is a balance between the factors that favour invariance and those that make for change, the new norms of behaviour that adolescents adopt are not too different from those of their parents, since in the majority of cases they are drawn from the same or a related culture. The fact that the search for new ideals begins comparatively early also means that young people have a long period of time during which to compare these ideals with the traditions of their parents, since the whole process takes place while they are still living within the firm social framework of their families. Normally, therefore, the forces for change are never present alone and are never allowed to work unchecked. Even at the time when *Wanderlust* is at its strongest, one feels a twinge of homesickness, and the older one gets, the stronger this feeling becomes. The spirit of rebellion is most intense at the beginning and declines with the years; one becomes increasingly tolerant towards one's parents and their memory, and there is scarcely a normal man who does not think more highly of his father at sixty than he did at sixteen.

Chapter twelve

Symbols and language

12.1 *Consolidation of symbolic meaning*

In the above sections on phylogenetic and cultural ritualization (Chapter 10.5 and 10.6) we discussed a series of processes that resemble symbolic representations of actions and objects. The freely created symbols of culture resemble the quasi-symbolic representations of phyletic ritualization in so far as both originally are vague in connotation: they never stand for a sharply defined object or action but invariably for a whole range of objects and actions, and especially emotions — a complex in which the elements are intertwined and for which there is no simple definition. If one asks where, at the non-linguistic or pre-linguistic stage, one can find a fixed, traditional symbol for a sharply delineated entity, the only answer that occurs to me is that it exists where a group of people have become united by a particular cultural symbol.

Apart from such group symbols, only linguistic symbols appear to possess a clearly circumscribed meaning. These, however, are quite distinct from any other form of ritual symbolization or quasi-symbolization, for they symbolize processes in the central nervous system which are governed by a set of complex phylogenetically programmed laws — the laws of abstract thought. This strict confinement of symbolic meaning to one particular concept is thus of a completely different kind from what one might call the 'consolidation' of a group symbol.

12.2 *The symbols of the group*

Groups larger than those held together by ties of personal friendship owe their cohesion entirely to symbols which have been evolved as a result of cultural ritualization and are felt by all members of the group to be of value. They are as deserving of respect and affection as the most dearly loved of one's fellow men, and in particular they are worth defending against all dangers. We have discussed the transference of affection from beloved persons to the traditions for which they stand.

Our most elementary response to such group symbols, and probably also the first to make its appearance in human society, actually appears to be homologous to the group defence put up by chimpanzees. We spring to the defence of the symbols of our civilization today, with the same hackle-raising, chin-protruding and mindless response of collective aggression with which a chimpanzee risks his life to defend his group. There is an old Ukrainian proverb that says: 'When the battle standard flies, the trumpet dominates the mind'.

The first symbols that our ancestors evolved for concrete realities, and perhaps the earliest symbols of any kind, may well stem from this group unity in time of war — war dance and war paint, for example, or flags. We know all too well how easily collective militant enthusiasm can become an all-destroying lethal factor.

12.3 *Verbal symbolization*

The only symbolizations which correspond to clearly defined concepts are, as far as I can see, those of language. The development of verbal language presupposes not only the processes of conceptual thought but also those of ritualization, which codify the symbol and make it a part of tradition. Syntactical language is a product of the fusion of the above two processes, born of that 'creative flash' which generated the power to represent, by an unambiguous linguistic symbol, a conceptually definable thought process. To realize how complicated a logical process symbolized in a single word can be, one should try to define — without using synonyms — the meaning of words like 'albeit' or 'yet'. The important fact that such words are translatable has already been discussed.

Even though the processes of cultural ritualization, which

themselves derive from the existence of a tradition, are probably far older than syntactical language, the latter is *the* vehicle of tradition *par excellence* and one of the most vital factors in the preservation of cultural stability. At the same time it is the most important organ of abstract thought and of human exploratory behaviour which, at its highest level, becomes research, and this makes it a key factor in the demolition and subsequent reconstruction of cultural structures. We discussed in Chapter 9 the sophisticated and highly complex innate programme that controls the ontogenetic development of language. The most important feature of this programme is the predisposition to attach a freely chosen symbol to each concept — in other words, to give names to objects and actions. From the observations made by Anne Sullivan, which I have described in detail above, it is clear that names first become applied to a complex consisting of an object and the activity connected with it, e.g. drinking milk or playing ball. But it is equally clear that there is also present an innate programme for nouns and verbs, making it possible to symbolize an activity independently of the object to which that activity is directed.

There are two further facts about the story of Helen Keller which should be stressed again here. The first is that at a particular stage in his childhood man has an overwhelming urge to put names to things, and feels a great sense of satisfaction when he succeeds in doing so. The second is that, notwithstanding the strength of this urge, he does not attempt to invent his own names, like Adam in the book of Genesis, but knows instinctively that they have to be learned, i.e. acquired from contact with tradition. The learning of language is therefore based on a phylogenetic programme which ensures that the child's innate power of abstract thought is integrated and reintegrated on every occasion with the vocabulary that belongs to its cultural tradition. This is a miracle of creation which cannot fail to fill us with a sense of awe and reverence every time we have the opportunity of witnessing its repetition in a young child.

Chapter thirteen

The aimlessness of cultural development

13.1 *Predetermination and free will*

One meets with passionate opposition from non-biologists when one attempts to explain that, in spite of its universal tendency to develop from the simple to the complex, from the probable to the improbable — in a word, from the lower to the higher — the whole of organic life is governed by the laws of chance and necessity. The widespread rejection of Jacques Monod's famous work, for instance, is emotional in origin and can only be understood as such. To realize that the great laws of nature admit of no exceptions seems to conflict with our consciousness of free will and with the value we attach to it as one of the supreme human possessions and an inalienable human right. The experiential fact that we possess free will, and our scientific awareness that our actions are physiologically determined, constitute a paradox to which I hope to offer a solution in my second volume.

It is almost equally difficult to accept the notion that the evolution of our culture is not controlled by our will, still less by our powers of abstract thought and reason. Right down to modern times most philosophers of history have clung to the conviction that man's historical development, the rise and fall of successive cultures, has followed a pre-existing plan and is governed by an underlying principle.

13.2 *Evolutionist approach to culture*

Like any species of animals or plants, and in fact like any business undertaking, every culture develops independently, at its own risk and by its own means, in a historically unique, unprecedented and unrepeatable process. As explained in Chapter 9 (p.177 ff.), neither cultural nor phyletic evolution ever follows a teleologically predetermined plan. The philosophers of history were extremely late in recognizing these facts: to the best of my knowledge Arnold Toynbee was the first to see them.

If we attempt to apply the approach and methods of causal analysis to the investigation of cultural development with the utopian aim of predicting and directing its course, we must begin by realizing that the factors which cause the increase of knowledge within any particular culture are analogous to those which govern the evolution of species. The way of regarding culture as a living system and of investigating the factors favouring stability, as well as those working for change, may enable us to interpret the development of a culture in terms of natural science.

I particularly wish to demonstrate that the cognitive function of culture, the acquisition and accumulation of knowledge, emerges through processes that are parallel to those which occur in phylogenetic development. This is all the more remarkable because it is the reason and intelligence of individuals that furnish all the information which becomes absorbed into the store of transmitted, supra-individual knowledge. In a mysterious and somewhat uncanny way a culture swallows up all the efforts of the individuals within it and turns them into a form of 'general knowledge', a sort of public opinion on what is true and real. The process is most admirably analysed in the book *The Social Construction of Reality* by Peter Berger and Thomas Luckmann, which I have already mentioned. But also involved in this process are many other factors that I have not mentioned, which have one thing in common — namely, that they are not consciously or rationally controlled. Thus the public opinion that prevails in a culture bears a closer resemblance to the body of information available to, and the consequent adaptedness of, an animal species, than to the knowledge which an individual possesses and is able rationally to apply. Although, in the brief space of a few centuries, modern science as a new form of collective cognitive endeavour has been forcing human thought into a strictly organized system of objectivity and, therewith, of consensus, the 'public opinion' thus brought about is by no means free of the non-rational influences emanating from the culture as a whole. The scientist too is a child of his age and of his culture.

The complete lack of rational planning in the development of culture and its products shows itself most surprisingly just where one would expect to find such planning — when, for instance, an engineer sits down at his drawing board to design, say, a railroad car. One would hardly credit that his procedures could be so strongly influenced by the forces of irrationality that contribute to cultural invariance (see pp.169 ff.). But if we look at the stages in the development of the railroad coach over the past century or so, we can see how remarkably men cling to tradition. It is almost as though one were confronting the outcome of a process of phylogenetic differentiation.

To start with, a horse carriage was simply put on iron wheels. Then the wheelbase was found to be too short, so it was lengthened, and with it the carriage itself. Instead of now constructing an independent form of bodywork suited to the longer carriage, the engineers designed a series of identical horse carriages of the traditional shape and strung them together. Each such carriage was joined to the next, producing compartments, but the doors, each with a large window, remained the same, together with the smaller windows on either side. Since the partitions between the compartments also remained in place, the conductor had to hold on to the handles and fight his way along the outside running board that stretched the length of the train. To keep to what has been tried and tested, and to be reluctant to embark on something entirely new, is characteristic of irrational thought, yet nowhere is it more in evidence than in those products of technology in which it prevents obvious solutions to problems from being found (see Fig.3).

Clearer, if less surprising, examples of our tendency towards an irrational conservatism in such matters are afforded by objects whose form is less dependent on function and therefore more susceptible to pressure from other quarters, such as its symbolic significance in some kind of ritual. In his book *Kultur und Verhaltensforschung* Otto Koenig made a comparative study of the historical development of military uniforms. He showed that not only could the concepts of homology and analogy be applied without qualification to this subject, but that the phenomena of functional change, reduction and vestigialization are also present

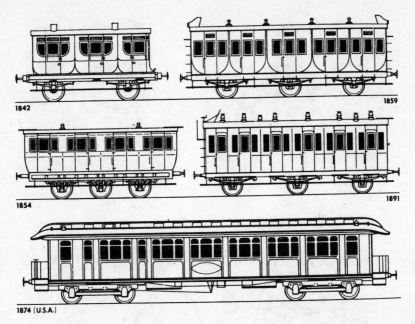

Figure 3 Example of the survival of non-functional elements in technical development. The original form of the horse carriage is retained. Even in the American coach, with its centre aisle, the windows are arranged in the same way as in the carriage, though the doors have been done away with. The development of the aisle came earlier in the United States than in Europe.

exactly as in phylogeny. One particular case (see Fig. 4) is that of the development of the piece of armour known as the hauberk, which, starting as a functional chain mail protection for the neck, eventually became an insignia of rank.

There was no plan for the development of these cultural products. They served specific functions, like organs, and the parallels between their historical development and the phylogenetic evolution of organs make one suspect that analogous forces are at work — in particular, that it is natural selection and not rational planning which is the dominant factor.

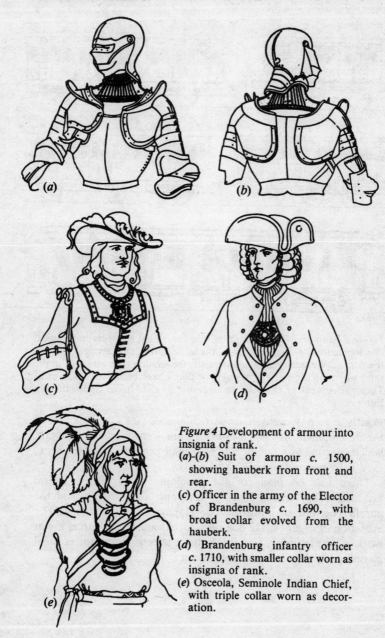

Figure 4 Development of armour into insignia of rank.
(a)-(b) Suit of armour c. 1500, showing hauberk from front and rear.
(c) Officer in the army of the Elector of Brandenburg c. 1690, with broad collar evolved from the hauberk.
(d) Brandenburg infantry officer c. 1710, with smaller collar worn as insignia of rank.
(e) Osceola, Seminole Indian Chief, with triple collar worn as decoration.

Chapter fourteen

Oscillation and fluctuation as cognitive functions

14.1 *Physical and physiological oscillation*

Any self-regulating process in whose mechanisms inertia plays a role tends towards oscillation. If a compass is shaken, the needle will swing to and fro several times before coming to rest in its 'proper' position, and it is difficult to construct regulators of any sort, such as thermostats, which, when compensating for a disturbance, do not first overshoot the 'value of reference' before returning to it in a dampened oscillation. This is even truer in the physiological field. The level of the so-called rest excitation of a neural element drops to zero after the element has discharged an impulse — that is, to a state of absolute non-excitability — after which it does not return to its former value on a dampened asymptotic curve, but overshoots it considerably. This is why the 'refractory period', as neurophysiologists call it, is followed by one of increased excitability and the previous value of rest excitation is only regained after a number of oscillations.

14.2 *Pseudo-topotaxis*

In his book *Die Orientierung der Tiere im Raum*, Alfred Kühn described a form of orientation in which an oscillatory process is used to produce a cognitive function. The sole component of this orientation mechanism is the phobic response, yet it is capable of locating the precise direction of its goal as efficiently as the topotaxis described on p.51 f. Kühn therefore called it 'pseudo-topotaxis'.

A good example is the way in which the marine snail, *Nassa*, locates its prey. Aroused by the scent, it emerges from the sand and waves its long breathing tube, or siphon, from side to side in its search for the source of the stimulus gradient and crawls forward in a random direction. The turn, controlled by the direction of the stimulus, characteristic of all topic responses, or taxes, is thus absent (see p.52), and the snail continues in the same haphazard direction until the differences in the concentration of the scent stimulus that it picks up through the lateral movements of its siphon begin to decrease. These differences will naturally be at their greatest at the moment the snail crosses at right angles the direction that leads to its goal, at which moment its siphon moves to and fro along a line pointing precisely in that direction. But instead of now turning at right angles, as one would expect, the animal makes an acute-angled turn — a phobic response of the type described earlier (p.51). By constantly repeating this process, it gradually works its way in zig-zag fashion towards the source of the scent until its siphon finally makes contact with its prey, whereupon it 'pounces'.

14.3 *Oscillation between hyperthymic and hypothymic moods*

As we all know from self-observation, a person's moods oscillate between states of elation and states of depression. Particularly in the case of creative thinkers and artists we find periods of inspiration and activity alternating with spells of discontent and sloth, both of varying duration. The pathological exaggeration of such fluctuations is called a manic-depressive state, in which the intensity and duration of such moods reach dangerous proportions.

There can be any number of intermediate stages between a 'normal' and a 'pathological' fluctuation, between hyperthymic and hypothymic states — mania and melancholia, as psychiatrists used to call them. The relative duration of any of these two states also varies: a man who is envied for always appearing cheerful may well have to pay for his cheerfulness with intense, if short, bouts of depression which few know about. In contrast, a man may be generally of a melancholic disposition yet enjoy by way of compensation spells of concentrated activity and productivity. I use the words 'pay' and 'compensation' deliberately, because I am convinced that there is indeed a physiological relation between the two peaks of this oscillation.

'Normal' and necessary fluctuations of mood related to the time of day can be observed in virtually all healthy people. When, as regularly happens, I wake up in the early hours of the morning my mind is dominated by unpleasant thoughts. Perhaps I suddenly remember an important letter I should have written, or think that I ought not to have put up with the way somebody behaved towards me, or find things wrong with what I had written the day before. In particular I discover all kinds of dangers of which I was quite unaware before and which I must take steps to cope with. So intensely do I experience these things that I sometimes take a piece of paper and write them down. Feeling relieved, I then go to sleep again. When the time comes to get up, things look a lot less black, and I find myself reassuringly able to deal with my problems, which, I must emphasize, are invariably quite real and not products of my imagination.

Such fluctuations of mood, to which most of us are subject at one time or another, are probably the effects of a self-regulating cycle incorporating a moment of inertia, which produces oscillatory movement. We all know that when a source of depression is suddenly removed, we experience a surge of happiness, and vice versa. I therefore see the oscillation between hyperthymic and hypothymic states as a biologically necessary process which reflects a search, on the one hand for potential dangers, and on the other for opportunities that we can exploit to our own advantage.

When my friend, the late Ronald Hargreaves, a truly inspired psychiatrist, asked me in one of his last letters what possible survival value could be attributed to states of depressive anxiety, I replied that if my wife had not been rather prone to such moods, two of my children would no longer be alive. She is a doctor, and her anxiety had led her to diagnose a particularly dangerous form of tubercular infection in the children at an early stage, when all the doctors she had consulted pooh-poohed the very idea. Had she not acted on her own diagnosis the children would have died.

External forces can greatly influence the amplitudes of these fluctuations of mood, and this too has its survival value. If a man is told that he has lost his job, or that he is suffering from diabetes, it seems logical that a mechanism should be set in motion which causes him to become acutely alert to the new dangers which are sure to arise from his changed situation. Conversely, when a man recovers from sickness or wins first prize in a sweepstake, it is very sensible to look for ways to profit from his good fortune.

Our passivity in times of depression also has a teleonomic purpose. When wild animals are on the look-out for foes, they remain motionless, using only their sense organs to try and detect the presence of an enemy. Nor is it by any means only the weakest or most timorous members of the animal kingdom that have to be alert to danger. The gander that is assigned the task of keeping watch while the flock is peacefully feeding is the strongest and bravest of the older males. *'Videant consules ...'* ('Let the consuls see to it ...') was a sound Roman adage.

The same applies to the physical activity that accompanies our states of elation. It is chiefly our receptor mechanisms that are needed to detect danger, whereas motor activity is required when it is a question of taking advantage of new opportunities.

The reciprocal oscillations of the threshold levels of all combinations of stimuli that elicit alternate hyperthymic and hypothymic states serve the purpose of what is known in cybernetics as a 'scanning mechanism'. This mechanism fulfils the dual role of guarding against potential dangers and looking for new opportunities to exploit. It thus clearly performs a cognitive function.

14.4 *The fluctuation of public opinion*

As has already been said, the public idea of what is true and real is based on a highly complicated system of social interactions. Since processes having a definite degree of inertia play a part in these, public opinion tends to oscillate. It was Thomas Huxley who said that every new truth starts as a heresy and ends up as an orthodoxy. If one were to interpret the word orthodoxy as signifying a body of rigid and mindless dogma, this would be a depressing statement, but if we take it as meaning the body of moderate views subscribed to by the majority of people in a given culture, we can see that what Huxley is describing is a characteristic cognitive function of human society.

The significance of genuinely new, epoch-making discoveries, especially in the natural sciences, is almost invariably overrated at the beginning, and by the discoverer himself more than anyone else. It is the prerogative of the genius who has found a new explanatory principle to over-assess its scope. Jacques Loeb thought he could explain all animal and human behaviour in terms of the principle of tropism; Pavlov thought he could do so on the basis of the conditioned reflex, while Freud was guilty of some comparable errors. One great scientist who *underrated* the importance of his discovery was Darwin.

Even within the confines of a particular school of thought, the formation of a new common view begins with someone departing from what had hitherto been accepted, but overstating his case. His colleagues, or more desirably his pupils, less inspired but with greater powers of analysis, then have the task of damping down the excessive oscillation and arresting the movement at the proper point. The reverse process can actually hinder the advancement of knowledge by establishing a dogma. For if the discoverer of a new truth finds himself in the company not of critical pupils and colleagues but of faithful disciples he may become the centre of a new religion, and although this may bring great blessings to mankind in general, it is not a desirable state of affairs in science. The discoveries of Freud have been misused and distorted in this way.

In a civilization as a whole, in substantial cultural groups, and especially in the natural sciences, the process of oscillation or fluctuation that follows each new act of discovery fulfils a cognitive function of a special kind. In the first instance the community tends to overestimate the new discovery in much the same way as does the discoverer himself. This collective over-assessment reaches excessive degrees if the discovery becomes fashionable. It is especially a newly discovered *method* that tends to become a fashion. And if these fashionable new methods involve considerable expense, such as the use of computers, which is so widespread today, they become status symbols in the eyes of young scientists.

Such exaggerated attitudes may have a positive value inasmuch as attempts to apply a new hypothesis or a new methodology to every situation, whether appropriate or not, often reveal an unexpected applicability of the new principle which more cautious procedures might not have uncovered. Conversely, if a new principle is uncritically adopted and applied, it may produce unintended and absurd results which prejudice public opinion against it and may eventually cause it to be consigned to oblivion. There are many examples of this in the history of science. On the other hand, the wrong-headed opposition to Darwin's theories, caused by a revulsion in public opinion, provoked new research and led to the discovery of powerful new arguments in their favour.

Viewed as a whole, this fluctuation of public attitudes thus produces a situation in which people with strongly held, diverging opinions seek arguments for or against a new theory, its opponents contributing in their own way towards giving it a firm foundation and determining its scope. This process functions like a scanning mechanism, and the point at which opinion finally stabilizes is closer to the truth than the position taken up by the discoverer in his first flush of excitement and success.

As long as the arguments on both sides remain motivated by a pure desire for truth, the fluctuations will continue damped down and will eventually come to rest at the point of intelligent consensus. But as soon as other urges and instincts become involved there is a risk that the differences between the rival parties will bring about the creation of two opposed ethnic groups, each convinced of the rightness of its own opinion, and roused, in its defence, to that kind of collective militant enthusiasm, the deadly dangers of which I have tried to describe in detail in my book *On Aggression*.

In such a situation the hypothalamus gets the better of the cortex, and the opposing opinions begin to lose their content of truth. The process is accelerated by the attempts of both sides to gain supporters by reducing their creeds to a series of short and simple statements, preferably slogans designed to be chanted in chorus. This increasingly stultifies the opposing opinions and correspondingly renders them less and less mutually acceptable. The final result is a self-generating form of oscillation which can only end in catastrophe.

Chapter fifteen
Behind the mirror

15.1 *Retrospect*

In the foregoing pages I have made a bold, maybe overly bold, attempt to give a survey of man's cognitive mechanisms. The real justification for such an undertaking can only derive from the possession of a comprehensive knowledge to which I cannot lay claim. But I felt entitled to embark on the task partly because no one else has hitherto ventured to do so, and partly because we urgently need such a comprehensive reflection on the various cognitive functions which constitute what Popper calls our perceiving apparatus. Each of these functions is apt to miscarry and to furnish distorted images of the reality around us, and this is especially true of the processes by which the collective knowledge of our culture is acquired and stored. We cannot understand pathological conditions or attempt to cure them without first knowing what the physiological process ought to be; at the same time a pathological state often provides us with the key to understanding the nature of the normal healthy function.

Readers will have observed that, particularly in later chapters, I have often advanced hypotheses about the physiological nature of cognitive functions on the basis of malfunctions of the mechanism in question. These cases will be discussed in a later volume, but I could not help anticipating this discussion at times.

Many, perhaps most, disorders of social conduct, whether this conduct be innate or the product of cultural norms, induce in normal men strong feelings of aversion and rejection. The problems of value and of the disturbances in the social behaviour of our fellow men which elicit our negative valuation are of course closely connected with the subject of this book, so that my firm decision to omit them entailed some rather abrupt and arbitrary cuts. None the less, it seemed to be advisable to concentrate first on the workings of our cognitive faculties.

As I stated in the Prolegomena, my approach to this, as to all biological processes, is based on that epistemological attitude which Donald Campbell called 'hypothetical realism'. The fact that throughout our investigations we have nowhere found anything to contradict the hypothesis on which our realism is founded, makes me feel that my book serves to strengthen this hypothesis to some extent.

15.2 *The importance of a scientific approach to cognition*

Instinctive — that is, phylogenetically programmed — norms of behaviour which resist all cultural influences play an important part in human social behaviour. Our statement of this indubitable fact is often interpreted as the profession of extreme cultural pessimism. Of course this is an error. To call attention to certain dangers is to demonstrate that one is in fact *not* a fatalist!

It was only for the sake of simplicity that in the foregoing pages I have kept up the fiction that most people are unaware of the processes of cultural evolution and decay, or at least that what we do know about the subjects I have been discussing does not and can not affect the course of history. This may have given the impression that I was claiming to be the only one who is propounding the necessity of investigating man's cognitive apparatus — the mechanism 'behind the mirror' — in terms of natural science. Nothing could be further from my mind than such arrogance. On the contrary, I am deeply aware that the epistemological and ethical views I have advanced here are shared by a rapidly growing number of thinkers. One sometimes finds that at a given moment certain discoveries are 'in the air'.

There are in my view definite signs that a self-recognition of all cultural humanity, a collective self-knowledge derived from natural science, is beginning to spring up. If, as is entirely possible, this movement grows, the intellectual aspirations and energies of mankind will be raised to a higher level of integration, as in the distant past the 'creative flash' of reflection and meditation raised man's powers of understanding to a new and higher level. A *reflecting self-investigation* of a culture has never yet come into being on this planet, just as objectivating science did not exist before the time of Galileo.

The scientific investigation of the structure of human society and its intellectual processes is a task of mammoth proportions. Society is the most complex of all living systems on earth, and our knowledge of it has barely scratched the surface. Yet I believe that man stands at a turning point in history and has at this moment the potential capacity to scale new and unknown heights.

To be sure, mankind also finds itself in greater danger than ever before. But the modes of thought that belong to the realm of natural science have, for the first time in world history, given us the power to ward off the forces that have destroyed all earlier civilizations.

Supplementary notes

I am indebted to my friends, in particular Eduard Baumgarten, Donald MacKay, Otto Rössler and Hans Rössner, for their observations and comments on the manuscript of this work. The following supplementary notes are based on these comments.

[1] (p.16) Some time ago there was a cartoon in *The New Yorker* of a giant computer which had just produced a print-out reading *'Cogito, ergo sum'*. The Cartesian *Cogito*, however, rests on a process not of knowing but of doubting — doubting, in the first place, the validity of subjective experience as a basis for a theory of knowledge. From my standpoint of 'hypothetical realism' it is pointless to deny the patent reality of subjective experience. This primary experience, the initial datum, so to speak, is not only the basis of phenomenology and a large part of philosophy in general but also the source of all our indirect, transmitted knowledge about the material world around us — 'distal knowing', as Campbell called it, borrowing an image from anatomy. Many would-be objective scientists tend to forget this.

The fact that I regard our knowledge of our own experience as immediate and incontrovertible must not be interpreted to mean that all other, indirectly acquired, 'distal' knowledge is uncertain, as non-realist philosophers insist in their various ways. The only thoroughly logical non-realist philosophical position is that of solipsism.

[2] (p.16) This criticism of Goethe's idealism may seem at odds with my choice of the couplet from the *Zahme Xenien* which I have used as the motto for this book. There is another version of the couplet in the Introduction to Goethe's *Theory of Colours*, where we also find the following: 'The eye owes its existence to light. From insignificant, secondary physical organs light has created an organ worthy of itself, and the eye was formed from light and for light, that the inner light might rise to meet the light from without'.

What a remarkable mixture of inspired insight and total misinterpretation of the facts! Goethe the visionary saw with unerring clarity that an organism and its environment are matched. But Goethe the typologist, close as he seemed to come to it on occasions, could not grasp the nature of the process by which they are matched. We thus find ourselves facing the extraordinary paradox that, for all his profound understanding of the dynamic forces of nature, he remained blind to the fact that the creative force is life itself, and that it is not the reflection of some pre-established harmony when the radiant inner light 'rises to meet the light from without'.

We know that the eye developed as a result of progressive adaptation to the properties of light. But there are also correspondences which cannot immediately be said to have been brought about by a process of adaptation. Thus men hold certain values which bear an obvious relationship to the whole evolutionary process of the organic world, but to explain this correspondence in purely evolutionist terms would be forced and unconvincing. The source of the correspondence lies in the fact that every normal man regards the natural function of organic evolution to produce order out of disorder, the unusual out of the predictable, as an achievement of the highest rank, a process that establishes standards.

The scale of values that we use, stretching from 'high' to 'low', is equally applicable to animal species, cultures and works of art. This correspondence might have arisen because man, as a thinking being, has become aware of certain processes that are active in the whole of the organic world; in this respect feeling, thought and being would thus be one, and a value judgement based on feeling would be *a priori*, i.e. innate and necessary. Alternatively one might explain the correspondence by assuming the presence in man of an innate releasing mechanism which makes him see order as intrinsically preferable to chaos — but this hypothesis, as I have said, strikes me as somewhat far-fetched.

We must emphasize that there is no connection between accepting the *a priori* nature of value judgements and believing in supranatural, vitalist forces. If value judgements are indeed *a priori*, then the pure work of art, created by man without giving thought to its 'relevance' or practical use, may be regarded as the material expression of these values and thus as an allegory of the process of creation, for the creative principle does not reside in the perishable stone of the statue or in the canvas and paint of the picture, but in the artist who created these works. True, a living organism perishes, as does a work of art, but an organism is not a creature of the artist: it is the product of its own in-dwelling power, of an immanent creator, so to speak. A living creature is not a symbol or an allegory, 'standing for' something else, but conscious reality in its own terms. To see it merely as a symbol of the imperishable and the eternal is the source of the erroneous anthropomorphic conception of creation discussed on p.16.

[3] (p.18) Readers may be surprised to find medical science suddenly drawn into the context of comparative ethology. The two are, however, closely linked. Comparative ethology originated when malfunctions of innate behaviour patterns showed that these patterns were physiological in nature. For instance, if one observes an animal performing an innate behaviour pattern in natural conditions, such as a wolf burying a piece of its prey in a safe place, one learns nothing about the physiology of its action. If, on the other hand, one watches a young dachshund take its bone into a corner of the living room, try to carry out the motions of digging a hole, put the bone down at the spot where it had been trying to dig and finally shovel invisible earth with its nose in order to carefully conceal the bone, one realizes that the animal's entire behaviour pattern is innate and not influenced in any way by external stimuli.

In basically the same manner pathological phenomena provide the most important source of our information about normal physiological conditions, and the reciprocal relationship between physiology and pathology in general also obtains in that between the physiology and pathology of behaviour — that is to say, between comparative ethology and psychiatry.

[4] (p.129) One may note that Chomsky frequently refers to

Wilhelm von Humboldt's work on language. In his treatise *Über die Verschiedenheiten des menschlichen Sprachbaues* of 1827 Humboldt writes:

> Language is, in its essence, not an achievement or deed (*ergon*) but an activity (*energeia*). Its true definition can therefore only be genetic. It is the ceaseless effort of the mind to make articulated sound capable of expressing thought ... The constancy and consistency of this mental effort, conceived in the totality of its context and systematically deployed, is what constitutes the form of language.

And elsewhere:

> Language is the formative organ of thought. Intellectual activity, something totally interior that passes almost without trace, is made exterior in speech through sound and becomes accessible to the senses, also receiving permanent form through writing. All spoken sounds and written symbols are produced in this way. Language is the embodiment of the sounds thus uttered, and of the combinations and transformations of these sounds according to laws, customs and analogies proper to intellectual activity and the system of sounds that it has evolved. These sounds, combinations and transformations are themselves contained in the *corpus* of all spoken or written language. Mental activity and language are therefore one and inseparable: it is not even possible to say that the former is the producer and the latter the product. For although everything that is spoken is a product of the mind, it is conditioned, as part of the pre-existing *corpus* of language, which lies outside the activity of the mind, by the laws and sounds of language itself, and thus, by becoming part of this *corpus*, reacts in turn upon the mind. Mental activity is compelled to enter into a union with sound, for in no other way can thought achieve clarity, or impression become concept. The mind produces and fashions sound of its own volition, turning it into articulate communication and thereby creating a body of sound which in turn exercises its own independent, guiding influence on the mind itself.

[5] (p.187) Humboldt's work on language (see note 4) may again be quoted here.

[6] (p.193) The direct comparison of animal species with human cultures tends to arouse the opposition of those with a highly developed sense of the differences between higher and lower living systems. The undeniable fact that cultures are highly complex intellectual systems, resting on a basis of symbols expressive of cultural values, causes us to forget, given, as we are, to thinking in terms of opposites, that they are natural structures which have evolved along natural lines (see Chapter 2 for the development of new system characteristics, and Chapter 3.4 on the fallacy of thinking in antitheses).

The different values that our feelings lead us to ascribe to systems on different levels of integration are not discussed here but will be left for a second volume, in which I shall also discuss the feelings of rejection and antipathy that we experience whenever the forward movement of evolution is reversed.

Bibliography

Baerends, G.P. Fortpflanzungsverhalten und Orientierung der Grabwespe (Ammophila campestris). *Tijdschrift voor Entomologie* 84 (1941) — Specializations in organs and movements with a releasing function. In: *Physiological Mechanisms in Animal Behaviour*. Cambridge 1950 (Cambridge Univ. Press).

Ball, W. & Tronick, E. Infant responses to impending collision: optical and real. *Science* 171, 818-20 (1971).

Bally, G. *Vom Ursprung und von den Grenzen der Freiheit. Eine Deutung des Spiels bei Mensch und Tier*. Basel 1945 (Birkhäuser).

Bateson, P.P.B. The characteristics and context of imprinting. *Biological Review* 41, 177-220 (1966).

Beach, F.A. *Hormones and Behavior*. New York 1948 (Cooper Square).

Bennett, J.G. *The Dramatic Universe*, 4 vols. Mystic, Conn. 1967 (Verry).

Berger, P.L. & Luckmann, Th. *The Social Construction of Reality*. New York 1966 (Doubleday).

Bertalanfy, L. v. *Theoretische Biologie*. Bern 1951 (Francke).

Bischof, N. Die biologischen Grundlagen des Inzesttabus. In: Reinert (ed.), *Bericht über den 27. Kongress der Deutschen Gesellschaft für Psychologie, Kiel*, Göttingen 1972 (Verlag für Psychologie). — Aristoteles, Galilei, Kurt Lewin — und die Folgen. *Zeitschrift für Sozialpsychologie*.

Bolk, L. *Das Problem der Menschwerdung*. Jena 1926.

Bower, T.G. The object in the world of the infant. *Scientific American* 225 (4), 30-8 (1971).

Bridgeman, P.W. Remarks on Niels Bohr's talk. *Daedalus*, Spring 1958.

Brun, E. Zur Psychologie der künstlichen Allianzkolonien bei den Ameisen. *Biologisches Zentralblatt* 32 (1912).

Brunswik, E. *The Conceptual Framework of Psychology*, Chicago 1952 (Chicago Univ. Press). — Scope and aspects of the cognitive problem. In

Bruner, T.S. et al. (eds.), *Contemporary Approaches to Cognition.* Cambridge, Mass. 1957 (Harvard Univ. Press).

Bühler, K. *Handbuch der Psychologie,* I. Teil. *Die Struktur der Wahrnehmung.* Jena 1922.

Butenandt, E. & Grüsser, O.J. The effect of stimulus area on the response of movement-detecting neurons in the frog's retina. *Pflügers Archiv für die gesamte Physiologie* 298, 283-93 (1968).

Campbell, D.T. Evolutionary epistemology. In: Schlipp, P.A. (ed.), *The Philosophy of Karl R. Popper.* LaSalle 1966 (Open Court Publ.) — *Pattern Matching as an Essential in Distal Knowing.* New York 1966 (Holt, Rinehart & Winston).

Chance, M.R.A. An interpretation of some agonistic postures; the role of "cut-off" acts and postures. *Symposia of the Zoological Society,* London 8, 71-89 (1962).

Chomsky, N. *Language and Mind.* Enl. ed. New York 1972 (Harcourt Brace Jovanovich).

Corti, W.R. Das Archiv für genetische Philosophie. Librarium. *Zeitschrift der Schweizer Bibliophilen Gesellschaft* 5, Heft I and II (1962).

Count, E.W. *Being and Becoming Human: Essays on the Biogram* (Behavioral Science Series) New York 1973 (Van Nostrand Reinhold).

Craig, W. Appetites and aversions as constituents of instincts. *Biological Bulletin* 34, 91-107 (1918).

Crane, J. Comparative biology of salticid spiders at Rancho Grande, Venezuela. IV: An analysis of display. *Zoologica* 34, 159-214 (1949). —. Basic patterns of display in fiddler crabs. *Zoologica* 42, 69-82 (1957).

Darwin, C. *Expression of the Emotions in Man and Animals.* (1872).

Decker, H. *Das Denken in Begriffen als Kriterium der Menschwerdung.*

Dethier, V.G. & Bodenstein. Hunger in the blowfly. *Zeitschrift für Tierpsychologie* 15, 129-40 (1958).

Eccles, J.C. *The Neurophysiological Basis of Mind: The Principles of Neurophysiology.* London 1953 (Oxford Univ. Press). — *Brain and Conscious Experience.* New York 1966 (Springer). — *Uniqueness of Man* (Roslansky, J.D., ed.), Amsterdam 1968 (North Holland).

Eibl-Eibesfeldt, I. Angeborenes und Erworbenes im Verhalten einiger Säuger. *Zeitschrift für Tierpsychologie* 20, 705-54 (1963). Die !Ko-Buschmann-Gesellschaft. *Gruppenbindung und Agressionskontrolle.* München 1972 (Piper). — Expressive behaviour of the deaf and blind born. In: Vine, I. (ed.), *Social Communication and Movement.* London 1973, 163-93 (Academic Press). *Love and Hate.* London 1972 (Methuen); New York 1972 (Holt, Rinehart & Winston).

Erikson, E.H. Ontogeny of ritualisation in man. *Philosophical Transactions of the Royal Society,* London 251 B, 337-49 (1966).

Foppa, C. *Lernen, Gedächtnis, Verhalten. Ergebnisse und Probleme der Lernpsychologie.* Köln 1966 (Kiepenheuer und Witsch).

Fraenkel, G.S. & Gunn, D.S. *The Orientation of Animals.* Oxford 1961 (Clarendon Press); New York 1961 (Peter Smith).

Freyer, H. *Schwelle der Zeiten.* Stuttgart 1965 (Deutsche Verlagsanstalt). —

Theorie des gegenwärtigen Zeitalters. Stuttgart [6] 1967 (Deutsche Verlaganstalt).

Garcia, J.A. & Koelling, R.A. A comparison of aversions induced by X rays, toxins and drugs in the rat. *Radiation Research Supplement* 7, 439-50 (1967).

Harlow, H.F. Primary affectional patterns in primates. *American Journal of Orthopsychiatry* 30 (1960). — *The Maternal and Infantile Affectional Patterns.* (1960).

Harlow, J.F., Harlow, M.K. & Meyer, D.R. Learning motivated by a manipulation drive. *Journal for Experimental Psychology* 40, 228-34 (1950).

Hartmann, M. *Allgemeine Biologie,* 4th ed. (1953). — *Die philosophischen Grundlagen der Naturwissenschaften.* Jena 1948; 2nd ed. 1959 (G. Fischer).

Hartmann, N. *Der Aufbau der realen Welt.* 3rd ed. Berlin 1964 (de Gruyter).

Hassenstein, B. Kybernetik und biologische Forschung. In: *Handbuch der Biologie* 1, 631-719. Frankfurt 1966 (Athaenaion).

Heilbrunn, L. v. *Grundzüge der allgemeinen Physiologie.* Berlin 1958 (Deutscher Verlag der Wissenschaften).

Heinroth, O. Beiträge zur Biologie, namentlich Ethologie und Psychologie der Antiden. *Verhandlungen des V. Internationalen Ornithologen-Kongresses,* Berlin 1910. — Reflektorische Bewegungen bei Vögeln. *Journal für Ornithologie* 66 (1918). — Über bestimmte Bewegungs-weisen der Wirbeltiere. *Sitzungsbericht der Gesellschaft der naturfor-schenden Freunde.* Berlin 1930.

Heinroth, O. and M. *Die Vögel Mitteleuropas.* Berlin-Lichterfelde, 1924-28 (Bermühler), reprint 1966-68.

Hess, E.H. Space perception in the chick. *Scientific American* 195, 71-80 (1956) — *Imprinting.* New York 1973 (van Nostrand).

Hinde, R.A. *Animal Behavior, a Synthesis of Ethology and Comparative Psychology.* New York 1972 (McGraw-Hill).

Holst, E. v. *Zur Verhaltenphysiologie bei Tieren und Menschen.* Vols I and II. München 1969/70 (Piper).

Hopp, G. *Evolution der Sprache und Vernunft.* Berlin 1970 (Springer); Frankfurt 1972 (Suhrkamp).

Hull, C.L. *Principles of Behavior.* New York 1943 (Appleton).

Huxley, J. The courtship habits of the great crested grebe (Podiceps cristatus); with an addition to the theory of sexual selection. *Proceedings of the Zoological Society,* London 35, 491-562 (1914).

Immelmann, J.: Prägungserscheinungen in der Gesangsentwicklung junger Zebrafinken. *Die Naturwissenschaften* 52, 169-170 (1965). — Zur Irreversibilität der Prägung, *Die Naturwissenschaften* 53, 209 (1966).

Itani, J. Die soziale Ordnung bei den japanischen Affen. In: *Tier* 6, 8-12 (1966).

Jander, R. Die Hauptentwicklungen der Lichtorientierung bei den tierischen Organismen. *Verhandlungen des Verbandes deutscher*

Biologen 3, 28-34 (1966).

Jennings, H.S. *The Behavior of the Lower Organisms*. New York 1906.

Kawai, M. Newly acquired pre-cultural behavior of the natural troops of Japanese monkeys on Koshima Island. *Primates* 6, 1-30 (1965).

Kawamura, S. The process of sub-cultural propagation among Japanese macaques. In: Southwick (ed.), *Primate Social Behavior*. New York 1963 (van Nostrand).

Keller, H. *The Story of My Life*. 1902.

Koehler. O. Die Ganzheitsbetrachtung in der modernen Biologie. *Verhandlungen der Königsberger Gelehrten Gesellschaft* (1933). — "Zählende" Vögel und vorsprachliches Denken. *Zoologische Anzeiger Supplement* 13, 129-38 (1949).

Köhler, W. *The Mentality of Apes*. Rev. 2nd ed. 1973 (Routledge & Kegan Paul).

Koenig, O. *Kultur und Verhaltensforschung. Einführung in die Kulturethologie*. München 1970 (Deutscher Taschenbuch-Verlag).

Konishe, M. Effects of deafening on song development in two species of juncos. *Condor* 66, 85-102 (1964). — Effects of deafening on song development of American robins and black-headed grosbeaks. *Zeitschrift für Tierpsychologie* 22, 584-99 (1965). — The role of auditory feedback in the control of vocalisation in the white-crowned sparrow. *Zeitschrift für Tierpsychologie* 22, 770-83 (1965).

Kruuk, H. *The Spotted Hyena*. Chicago 1972 (Chicago Univ. Press).

Kuenzer, P. Die Auslösung der Nachfolgereaktion bei erfahrungslosen Jungfischen von Nannacara anomala (Cichlidae). *Zeitschrift für Tierpsychologie* 25, 257-314 (1968).

Kühn, A. *Die Orientierung der Tiere im Raum*. Jena 1919 (G. Fischer).

Lashley, K.S. In search of the engram. In: *Symposia of the Society for Experimental Biology IV, Physiological Mechanisms in Animal Behaviour*. Cambridge 1950 (Cambridge Univ. Press).

Lawick-Goodall, H. and J. van. *Innocent Killers*. Boston 1972 (Houghton Mifflin Co.).

Lennenberg, E.G. *Biological Foundations of Language*. New York 1967 (Wiley).

Lettvin, Maturana, McCullock & Pitts. What the frog's eye tells the frog's brain. *Proceedings I.R.E.* 47, 1940-1951 (1959).

Leyhausen, P. Über die Funktion der relativen Stimmungshierarchie (dargestellt am Beispiel der phylogenetischen und ontogenetischen Entwicklung des Beutefangs von Raubtieren). *Zeitschrift für Tierpsychologie* 22, 412-94 (1965).

Lieb, J. Die Tropismen. *Handbuch der vergleichenden Physiologie* 4 (1913).

Lorenz, K. *On Aggression*. London 1966 (Methuen); New York 1966 (Harcourt Brace Jovanovich). *Evolution and Modification of Behavior*. Chicago 1965 (Chicago Univ. Press). *Studies in Animal and Human Behavior*, Vol. I, London 1970; Vol. II, London 1971 (Methuen); Cambridge, Mass. (Harvard Univ. Press). — Stammes- und kulturgeschichtliche Ritenbildung, *Mitteilungen der Max Planck Gesellschaft* 1, 3-30 and *Naturwissenschaftliche Rundschau* 19, 361-7. — *Civilized Man's Eight Deadly Sins.* London 1974 (Methuen); New York 1974 (Harcourt Brace Jovanovich).

Mackay, D.M. *Freedom of Action in a Mechanistic Universe.* Cambridge 1967 (Cambridge Univ. Press).

Maier, N.R.F. Reasoning in white rats. *Comparative Psychology Monograph* 6, 29 (1929) — Reasoning in humans: I. On direction. *Journal of Comparative Psychology* 10, 115-143 (1930).

Mayr, E. *Animal Species and Evolution.* Cambridge, Mass. 1963 (Harvard Univ. Press).

Metzger, W. *Psychologie.* 4th ed. Darmstadt 1953. 1968 (Steinkopff).

Metzner, P. Studien über die Bewegungsphysiologie niederer Organismen. *Die Naturwissenschaften* II (1923).

Meyer-Eppler, W. *Grundlagen und Anwendung der Informationstheorie.* Berlin 1959 (Springer).

Meyer-Holzapfel, M. Triebbedingte Ruhezustände als Ziel von Appetenzhandlungen. *Die Naturwissenschaften* 28, 273-80 (1940).

Mittelstaedt, H. Über den Beutefangmechanismus der Mantiden. *Zoologische Anzeiger Supplement* 16, 102-6 (1953). — Regelung in der Biologie. *Regelungstechnik* 2, 177-81 (1954). — Regelung und Steuerung bei der Orientierung von Lebewesen. *Regelungstechnik* 2, 226-32 (1954).

Monod, J. *Chance and Necessity: An Essay on the Natural Philosophy of Modern Biology.* New York 1971 (Knopf).

Nicolai, J. Zur Biologie und Ethologie des Gimpels. *Zeitschrift für Tierpsychologie* 13, 93-132 (1950).

Peckham, G.W. and E.G. Observations on sexual selection in spiders of the family Attidae. *Occasional Papers of the National History Society of Wisconsin.* Milwaukee 1889.

Pittendrigh, C.S. Perspectives in the study of biological clocks. In: *Perspectives in Marine Biology.* La Jolla 1958 (Scripps Inst. Oceanogr.).

Planck, M. Sinn und Grenzen der exakten Wissenschaft. *Die Naturwissenschaften* 30 (1942).

Plessner, H. *Philosophische Anthropologie.* Stuttgart 1970 (Fischer).

Polanyi, M. Life transcending physics and chemistry. *Chemical and Engineering News* (1976). — *Personal Knowledge: Towards a Post-Critical Philosophy.* Chicago 1958 (Chicago Univ. Press).

Popper, K.R. *The Logic of Scientific Discovery,* rev. ed. New York 1962 (Harper & Row). — *The Open Society and its Enemies.* London 1945; 5th rev. ed. 1960 (Princeton Univ. Press).

Porzig, W. *Das Wunder der Sprache.* München/Bern 1950, 5th ed. 1971 (Francke).

Reese, E.S. The behavioral mechanisms underlying shell selection by hermit crabs. *Behaviour* 21, 78-126 (1963). — A mechanism underlying selection or choice behaviour which is not based on previous experience. *American Zoology* 3, 508 (1963). — Shell use: an adaptation for emigration from the sea by the coconut crab. *Science* 161, 385-6 (1968).

Regen, J. Über die Orientierung des Grillenweibchens nach dem Stridulationsschall des Männchens. *Sitzungsbericht der Akademie der Wissenschaften, Wien*; Kongress 132 (1924).

Richter, C.P. *The Self-Selection of Diets. Essays in Biology.* Berkeley 1943 (Univ. of California Press).

Rössler, O.E. *Theoretische Biologie.* Lecture at the Max-Planck-Institut für Verhaltensphysiologie, Seewiesen (1966).

Rose, W. Versuchsfreie Beobachtungen des Verhaltens von Paramaecium aurelia. *Zeitschrift für Tierpsychologie* 21, 257-78 (1964).

Schein, W.M. On the irreversibility of imprinting. *Zeitschrift für Tierpsychologie* 20, 462-7 (1963).

Schleidt, M. Untersuchungen über die Auslösung des Kollerns beim Truthahn. *Zeitschrift für Tierpsychologie* 11, 417-35 (1954).

Schleidt, W.M. Reaktionen von Truthühnern auf fliegende Raubvögel und Versuche zur Analyse ihrer AAMs. *Zeitschrift für Tierpsychologie* 18, 534-60 (1961). — Wirkungen äusserer Faktoren auf das Verhalten. *Fortschritte der Zoologie* 16, 469-99 (1964).

Schleidt, W.M., Schleidt, M. & Magg, M. Störungen der Mutter-Kind-Beziehung bei Truthühnern durch Gehörverlust. *Behaviour* 16, 254-60 (1960).

Schutz, F. Sexuelle Prägung bei Anatiden. *Zeitschrift für Tierpsychologie* 22, 50-103 (1965). — Sexuelle Prägungserscheinungen bei Tieren. In: H. Giese (ed.), *Die Sexualität des Menschen. Handbuch der Medizinischen Sexualforschung* 1968, 284-317 (Enke).

Schwartzkopff, J. Vergleichende Physiologie des Gehörs und der Lautäusserungen. *Fortschritte der Zoologie* 15, 214-336 (1962).

Sedlmayr, H. *Gefahr und Hoffnung des technischen Zeitalters.* Salzburg 1940 (Otto Müller).

Seibt, U. Die beruhigende Wirkung der Partnernähe bei der monogamen Garnele Hymenocera picta. *Zeitschrift für Tierpsychologie* 33 (1973).

Sherrington, C.S. *The Integrative Action of the Nervous System.* New York (1906).

Skinner, B.F. Conditioning and extinction and their relation to drive. *Journal of General Psychology* 14, 296-317 (1936). — *The Behavior of Organisms.* New York 1938 (Appleton-Century-Crofts). — Reinforcement today. *American Psychologist* 13, 94-99 (1958). — *Beyond Freedom and Dignity.* New York 1971 (Knopf).

Snow, C.P. *The Two Cultures.* London 1959, 1963 (Cambridge Univ. Press).

Steiniger, F. Zur Soziologie und sonstigen Biologie der Wanderratte. *Zeitschrift für Tierpsychologie* 7, 356-79 (1950).

Storch, O. Erbmotorik und Erwerbsmotorik. *Akademischer Anzeiger der mathematisch-naturwissenschaftlichen Klasse der Österreichischen Akademie der Wissenschaften* 1, 1-23 (1949).

Taub, E., Ellman, St. J., & Berman, A.J. Deafferentiation in monkeys. Effects on conditioned grasp response. *Science* 151, 593-4 (1965).

Thorndike, E.L. *Animal Intelligence.* New York 1911 (Macmillan).

Thorpe, W.H. *Science, Man and Morals*. London 1965 (Methuen).

Thorpe, W.H. & Jones, F.H.W. Olfactory conditioning in a parasitic insect and its relation to the problem of host selection. *Proceedings of the Royal Society*, London B, 124, 56-81 (1937).

Tiger, L. & Fox, R. *The Imperial Animal*. New York, 1972 (Dell).

Tinbergen, N. Die Übersprungbewegung. *Zeitschrift für Tierpsychologie* 4, 1-40 (1940). — *The Study of Instinct*. London and New York, 1969 (Oxford Univ. Press).

Tolman, E.C. *Purposive Behavior in Animals and Men*. New York 1932 (Appleton-Century-Crofts).

Trumler, E. *Mit dem Hund auf Du*. München 1971 (Piper).

Uexküll, J. v. *Umwelt und Innenleben der Tiere*. Berlin 1909, 2nd ed. 1921.

Weidel, W. *Virus und Molekularbiologie*. Berlin, 2nd ed. 1964 (Springer).

Weiss, P.A. The living system: determinism stratified. In: Koestler & Smythies (eds.), *Beyond Reductionism*. London 1969 (Hutchinson); New York 1970 (Macmillan). *Dynamics of Development: Experiments and Inferences*. New York 1968 (Academic Press).

Wells, M.J. *Brain and Behavior in Cephalopods*. London 1962 (Heinemann).

Whitman, Ch. O. *Animal Behavior*. 16th lecture from *Biological Lectures from the Marine Biological Laboratory*. Woods Hole, Mass. 1898.

Wickler, W. *Mimicry in Plants and Animals*. New York 1968 (McGraw-Hill).

Wundt, W. *Vorlesungen über die Menschen- und Tierseele*. Leipzig 1922 (Voss).

Wynne-Edwards, V.C. *Animal Dispersion in Relation to Social Behaviour*. London 1962 (Oliver & Boyd); New York 1962 (Hafner).

Zeeb, K. Zirkusdressur und Tierpsychology. *Mitteilungen der Nationalen Forschungsgeseilschaft Bern*, N.F. 21, 1964.

Index